COMPRESSORS AND EXPANDERS

CHEMICAL INDUSTRIES

A Series of Reference Books and Text Books

Consulting Editor
HEINZ HEINEMANN
Heinz Heinemann, Inc.,
Berkeley, California

Volume 1: Fluid Catalytic Cracking with Zeolite Catalysts,
Paul B. Venuto and E. Thomas Habib, Jr.

Volume 2: Ethylene: Keystone to the Petrochemical Industry,
Ludwig Kniel, Olaf Winter, and Karl Stork

Volume 3: The Chemistry and Technology of Petroleum,
James G. Speight

Volume 4: The Desulfurization of Heavy Oils and Residua,
James G. Speight

Volume 5: Catalysis of Organic Reactions,
edited by William R. Moser

Volume 6: Acetylene-Based Chemicals from Coal and Other Natural
Resources, *Robert J. Tedeschi*

Volume 7: Chemically Resistant Masonry,
Walter Lee Sheppard, Jr.

Volume 8: Compressors and Expanders: Selection and Application
for the Process Industry, *Heinz P. Bloch, Joseph A.*
Cameron, Frank M. Danowski, Jr., Ralph James, Jr.,
Judson S. Swearingen, and Marilyn E. Weightman

Additional Volumes in Preparation

COMPRESSORS AND EXPANDERS

Selection and Application for the Process Industry

Heinz P. Bloch
Exxon Chemical Americas
Baytown Olefins Plant
Baytown, Texas

Joseph A. Cameron
Frank M. Danowski, Jr.
Materials Engineering
Elliott Co.
Jeannette, Pennslyvania

Ralph James, Jr.
Exxon Chemical Co.
Florham Park, New Jersey

Judson S. Swearingen
Rotoflow Corp.
West Los Angeles, California

Marilyn E. Weightman
Materials Engineering
Elliott Co.
Jeannette, Pennsylvania

MARCEL DEKKER, INC.   New York and Basel

6654-0653

CHEMISTRY

Library of Congress Cataloging in Publication Data
Main entry under title:

Compressors and expanders.

   (Chemical industries ; v. 8)
   1. Compressors. 2. Expanders, Gas. I. Bloch,
Heinz P., [date]   . II. Series.
TJ990.C635 1982   660.2'83   82-13062
ISBN 0-8247-1854-2

These articles originally appeared in the <u>Encyclopedia of Chemical</u>
<u>Processing and Design</u>, Volume 10, edited by John J. McKetta, published
by Marcel Dekker, Inc., New York, 1979.

MARCEL DEKKER, INC.
270 Madison Avenue, New York, New York 10016

Current printing (last digit)
10 9 8 7 6 5 4 3 2 1

PRINTED IN THE UNITED STATES OF AMERICA

Preface

This book is an excellent and comprehensive presentation of current compressor technology. It also contains a section on turboexpanders. Compressors and turboexpanders are often used in sets, and much information is common to both, especially as to shaft couplings, thermal expansion and vibrations.

We have freely included easy-to-understand explanations. The introductory chapter explains various problems and limitations that are of concern to the designer and to the purchaser and user of compressors. The viewpoints and working information of manufacturers is clearly apparent throughout the book. We believe it will be especially useful to maintenance departments to aid in analyzing problems, also to engineers in anticipating possible operating advantages and problems.

Purchasers of compressors and turboexpanders should find this book a great aid and time-saver in reducing the number of possible choices for particular applications.

The field of compression requirements is enormous, and there are many types and variations of compressors. In this volume the preferred types for the various zones of the large spectrum of compression conditions are identified. The mechanical features, the available sizes, and their respective degrees of perfection are well presented in perspective.

Stress, vibration, corrosion and speed limitations have profound effects on this overall picture — more than usually is apparent. To aid in making this clear we have included numerous simplified descriptions of these effects in the Introductory Chapter.

Turboexpander types and applications are similarly covered, commensurate with their applications, including simplified methods of evaluating their performance.

Most of the material is so fundamental that we believe it will be widely applicable for many years.

We greatly appreciate the help and suggestions by the staff of Marcel Dekker, Incorporated.

By Judson S. Swearingen

Contents

Contents

Introduction

This first chapter is intended to serve as an introductory review of a few areas of compressor technology to make the subsequent chapters easier to read and understand. Also, for the reader's convenience, it covers a few subjects as follows that are common to subsequent chapters:

1. Review of design principles of a centrifugal compressor,
2. Shaft critical speeds,
3. Torsional vibrations of rotating systems,
4. Structural vibrations,
5. Resonance damage to impellers,
6. Gear speed increasers,
7. High-speed shaft couplings,
8. Shaft seals,
9. Optimization,
10. References.

Centrifugal Compressor Principles of Operation

A single stage centrifugal compressor receives the inlet stream axially into the "eye" of the rotating impeller. Crudely conceived, it has radial blades covered with a shroud so that it comprises a sequence of radial passages. The impeller spins and slings the liquid outward into a collecting zone leading to the discharge.

Now introducing several refinements:

First, the inlet should have the blades curved into the leading direction so as to slice into the non-spinning inlet stream with minimum disturbance. Once these sliced streams reach the beginning of the respective channels through the rotor, the channels bend outwardly from an axial direction to an essentially radial direction. In some impellers the axial inlet stream mushrooms into the radial direction, or partially so, before encountering the blades.

The performance of the impeller is based on Euler's equation, which is a statement of the momentum effects upon the impeller equated to the torque or work load on the impeller. This equation states:

$$T = \frac{C_{2u}U_2 - C_{1u}U_1}{g} \tag{1}$$

where T = work per pound pumped, ft-lb.
C = fluid velocity, ft/sec.
U = rotor tangential velocity, ft/sec.
g = 32.17
Subscripts
1 = inlet
2 = discharge
r = radial direction
u = tangential direction
(+ is impeller's rotating direction)

Since the incoming stream into the inlet normally is not whirling, it is apparent that C_{1u} equals 0. Thus, the second term of the equation vanishes, and

$$T = \frac{C_{2u}U_2}{g} \tag{2}$$

The discharge stream leaves the impeller rotating or spinning (in the same direction as the impeller) so it contains some kinetic energy. This kinetic energy $= \frac{C_{2u}^2}{g}$. (For convenience, it is assumed that the inlet axial velocity is equal to the radial component of the discharge velocity, C_{2r}, so as to eliminate concern about velocity in the flowing direction).

Deducting the kinetic energy from the total work (2) gives the pressure head gain across the impeller:

$$\text{Pressure head gain in impeller} = \frac{C_{2u}U_2}{g} - \frac{C_{2u}^2}{2g} \tag{3}$$

The purpose of the compressor is to increase the pressure head. Half or more of the head is already gained in the impeller, but the remainder of the energy is in the form of kinetic energy, and must be converted to pressure. This is accomplished in a "diffuser," characteristically represented by a sequence of blades around the periphery of the impeller which cut into the impeller discharge spinning stream of fluid with minimum disturbance, and then, as these separate streams flow through the passages between the blades of the diffuser, they are slowed down and thereby most of their velocity energy is converted into pressure head. This conversion takes place because these passages uniformly and gradually diverge.

For a sample calculation of compressor head gain, let's assume

  a) that 6% of the work input (T) into the impeller is spent in fixed losses (disk friction and seal leakage), leaving $C_{2u}U_2 = 0.94T$,

  b) that 9% of this remaining 94% of the power, T, is lost in friction and turbulence within the impeller passages, so $0.91\ C_{2u}U_2$ is the energy remaining in the impeller discharge stream,

  c) that the diffuser is 70% efficient in converting the kinetic energy into pressure energy, and

  d) that the impeller discharge tangential velocity is $C_{2u} = 0.6U_2$.

From assumption a):

$$0.94 \times (\text{input work}) = \frac{C_{2u}U_2}{g} \tag{1}$$

From b) and according to Euler's Equation:

$$0.91 \times \frac{C_{2u}U_2}{g} = (\text{impeller pressure head}) + \frac{C_{2u}^{\ 2}}{2g} \tag{2}$$

$$\text{From c) the diffuser pressure head gain} = 0.7\frac{C_{2u}^{\ 2}}{2g} \tag{3}$$

Total head gain = impeller pressure head + diffuser pressure head

$$= 0.91\frac{C_{2u}U_2}{g} - \frac{C_{2u}^{\ 2}}{2g} + 0.7\frac{C_{2u}^{\ 2}}{2g} \tag{4}$$

From d) $C_{2u} = 0.6U_2$, so from (4)

$$\text{Total head} = 0.91 \times \frac{0.6U_2^2}{g} - \frac{0.6^2U_2^2}{2g} + \frac{0.7 \times 0.6^2U_2^2}{2g}$$

$$= [(2 \times 0.91 \times 0.6) - 0.6^2 + 0.7 \times 0.6^2]\frac{U_2^2}{2g}$$

$$= (1.092 - 0.36 + 0.252)\frac{U_2^2}{2g}$$

$$= 0.984\frac{U_2^2}{2g}$$

where 0.984 is the "pumping coefficient," which is the head (of one stage) in terms of $U_2^2/2g$.

Its value usually falls in the range of unity.

Now consider an ideal situation, where the blades are radial at the outlet of the impeller and there is a large number of blades so that the stream leaves the impeller at the impeller tip speed and there are no impeller losses. Euler's equation for impeller head gain is $\dfrac{C_{2u}U_2}{g} - \dfrac{C_{2u}{}^2}{2g}$. If $C_{2u} = U_2$, and further, if the diffuser is 100% efficient the pressure head would be $\dfrac{U_2^2}{g}$ or $\dfrac{2U_2^2}{2g}$ which is a pumping coefficient of 2.

However, in practice the stage is more efficient if the blades lean backward and the value of C_{2u} is substantially less than U_2. Further, the efficiency of the diffuser is substantially less than unity, and also there are flow friction and turbulence losses, so this all brings the pumping coefficient down to a range in the order of unity.

As to efficiency in the above example, the losses (simplified) are:

$$\text{Diffuser loss} = (1 - 0.7)\frac{C_{2u}{}^2}{2g}$$

(See assumption c) above).

evaluated in terms of T as follows,

$$(1 - 0.06)T = \frac{C_{2u}U_2}{g} \qquad\qquad\qquad \text{See (a) \& (1)}$$

$$C_{2u} = 0.6U_2 \quad \text{or} \qquad\qquad\qquad\qquad \text{See (d)}$$

$$U_2 = \frac{C_{2u}}{0.6} \text{ Substitute, get}$$

$$(1 - 0.06)T = \frac{C_{2u}}{g} \cdot \frac{C_{2u}}{0.6}$$

$$= \frac{C_{2u}{}^2}{0.6g}$$

$$\frac{C_{2u}{}^2}{2g} = \frac{(1 - 0.06)(0.06)T}{2}, \text{ so, substituting, get}$$

$$\text{Diffuser loss} = \frac{(1 - 0.7)(1 - 0.06)(0.6)}{2}T = 0.0846T$$

Seal and disk friction 0.06T
(See assumption a) above).
Impeller internal loss . . . . . . 0.09(1 - 0.06)T = 0.0846T
(See assumptions a) and b) above).   _____

$$\text{Total Losses} = 0.2292T$$

$$\text{Efficiency} = \frac{\text{Total Power} - \text{Total Losses}}{\text{Total Power}}$$

$$= \frac{T - 0.2292T}{T} = 77.08\%$$

Specific Speed

If the flow is small by comparison to the diameter of the impeller, the eye should best be small. On the other hand, if the flow is large, there must be a large inlet eye. With relatively large flow the total of the fixed losses, such as disk friction and seal leakage, prorated against the larger flow is a smaller fraction of the total power. For small flows (for the same impeller diameter and rotating speed), these losses become large in proportion. Thus, maximum efficiency favors a relatively high ratio of flow to size. At still larger flows, pressure drop, varying as the square of the flow, becomes excessive.

This relationship between flow and diameter, taking speed and pressure rise into account, is expressed in an equation called the "specific speed," which is

$$N_s = N \frac{\sqrt{c.f.s.}}{h^{3/4}}$$

where N_s = specific speed
N = rev. per min
h = head, ft.
c.f.s. = cubic ft. per sec. at inlet

The optimal values of the specific speed are of the order of 125. Below 75 the efficiency rather rapidly falls off. In the range of 250 to 400 the impeller takes the form of a screw or propeller.

Under certain circumstances it is convenient to use other expressions of capacity, head, and speed. However, it probably is safer to think in terms of pumping coefficient and specific speed.

Further, concerning the diffuser: The above explanation uses a vane-type diffuser. However, especially in high specific speed units, a "vaneless diffuser" is applicable and often more efficient, and gives good efficiency over a wider range of flow.

To understand the performance of a vaneless diffuser, consider a simple fluid vortex, a whirling body of fluid. If it is flowing from the inside outward, the pressure will rise due to the centrifugal force acting on it by virtue of its whirling. This pressure rise comes at the expense of the whirling velocity so the velocity falls off as the fluid moves radially outward. The relationship is that the velocity is inversely proportional to the radial position (neglecting friction).

Thus, if the discharge from the impeller enters an annular passageway and moves from the impeller discharge diameter to a point in the diffuser where the diameter is twice that of the impeller, at that point the fluid is spinning at roughly half the velocity and retains at that point one-fourth the kinetic energy. Three-fourths of the energy has been converted to pressure (except for the friction loss). At the discharge of this vortex there can still be a volute passageway to gather up the still moderately spinning liquid and at the volute discharge to gently diverge the stream to recover some additional head.

It is apparent that vaneless-type diffusers take more radial space than vane diffusers.

Summarizing, the vane diffuser is usually more efficient, especially in low specific speed applications and requires less radial space, but has narrower operating range.

There has always been large demand for higher head than an impeller conveniently produces, so there is temptation to use lower specific speed impellers, that is, increase the diameter of the impeller without decreasing the speed of flow. This increases the disk friction beyond that usually tolerable. The disk friction varies as the cube of the tip speed and as the square of the diameter.

The suggestion has been made to place a freely spinning disk in the space between the rotating impeller surface and the stationary wall. Then there would be only half the shear velocity on each side of the free disk, which means one-eighth of the friction loss on each side. There are two sides to the free disk so the disk friction would be one-fourth of its usual value. This idea was patented about 100 years ago, but it has never caught on. It is mechanically inconvenient to do, and furthermore, the use of suitable specific speeds has largely overcome the problem, since choosing a suitable specific speed is no longer difficult in most cases.

There is considerable discussion in the literature about the shapes of the passageways through the impeller. The flow through the impeller is complicated because of the curvature and the presence of accelerating forces and centrifugal forces. In general, it is helpful to have some extension of the eye to allow the impeller to smoothly receive the liquid between blades before turning radially outwards; and then to lean the blades backward for several reasons, mostly to improve the uniformity of the stream velocity as it leaves the impeller. This also reduces the tangential velocity C_{2u} at the impeller discharge and thus leaves less of the energy in the form of kinetic energy to be converted in the less efficient diffuser.

Shaft Critical Speeds

Reference is commonly made to the "critical speed" of a rotor shaft assembly rotating in supporting bearings. This concept usually presumes reasonably good support by the bearings, and the "critical speed" relates to the flexibility and weight distribution combination of the rotor. It is the speed at which any degree of shaft radial elastic distortion is producible by the centrifugal force that results from such distortion. The farther the shaft deflects, the greater the centrifugal force becomes so the force and the radial distortion match at all radial distortions. There always is some residual imbalance, and this imbalance causes the distortion or deflection from alignment to gradually increase according to a spiral, increasing without limit (unless some damping effects overcome it).

At a speed usually several-fold that of this "first critical" there is a "second critical." It is a similar phenomenon except that the shaft in this case assumes an "S" shape, having a point of inflection in its configuration. There are other higher shaft criticals possible at much higher speeds. There also are some modes of shaft radial vibrations largely involving bearing flexibility.

A journal bearing oil film, in general, has linear radial flexibility with load, and, in addition, the bearing support may be somewhat flexible. Such radial flexibility in combination with the kinetics of the rotating shaft assembly usually gives rise to modes of vibration other than the primary shaft critical described above. For example, a shaft mounted in two flexible bearings can rock in the bearings even though the shaft may be comparatively rigid. Usually the bearing oil films are flexible enough to permit this form of vibration, and it occurs at a speed well below the "first shaft critical." This rocking vibration in recent literature often is referred to as "the first critical."

Looking at it this way, if the radial vibration amplitude of the shaft were plotted against speed, there may be several peaks; the first of which may result from this rocking movement.

Another possible mode is a total rotor radial gyration in the two oil films.

If the bearings are not similar, or if the rotor is not symmetrical, there will be two frequencies for the rocking movement and perhaps other minor complications. Thus, a plot of the shaft radial deflection as the speed increases may show a number of peaks, and they are sometimes indiscriminately referred to as criticals, the first being that which occurs at lowest speed.

The average viscosity of the lubricant in the bearing film linearly affects the rigidity of the oil film. As the speed increases, so does the spring rate of the bearing oil film, provided the viscosity and clearance remain constant. However, as the speed increases the viscosity usually decreases due to increasing temperature, and sometimes the bearing clearance increases; thus, these various "criticals" are tuned to reasonably constant frequencies regardless of the speed of the shaft (after it has reached a reasonably high speed and considerable heat is being produced in the oil film).

Another phenomenon—that of "oil swirl," or "oil whip," or "half speed gyration" occurs at roughly twice the lowest of these resonant speeds, provided the flexibility of the shaft does not complicate the resonating system.

To explain this, envision a journal bearing with a centrifugal load of some kind—perhaps due to imbalance—distorting the oil film, making it thinner on one side and thicker on the other. The thick side usually is referred to as an "oil wedge." The oil film is dragged around at an average of half the shaft speed for obvious reasons, and so this oil wedge is dragged around at half the shaft speed. Thus, it imposes an exciting force at half the frequency of the rotating speed. If the rotating speed should be twice the resonant frequency of an oil film mode or any mode, then clearly such resonance would be excited by this half-speed exciting force due to the wedge being dragged around at half-speed. Such excitement of the resonance would increase the radial load which, in turn, increases the magnitude of the oil wedge, and makes the oscillation more and more violent.

Various methods of minimizing this "oil wedge" phenomenon are used such as step bearings, and so on. The most successful seems to be a pivoted shoe bearing, and its use significantly increases the speed where such phenomenon becomes objectionable.

In many, if not most rotating machines, especially multi-stage machines with a long relatively thin shaft, it is highly desirable to operate above one or more of these "criticals" (but not above oil swirl). In such assemblies it is desirable and, in fact, essential to limit the radial magnitude of these vibrations. The limit of the amplitude of vibration is at the point where the power absorbed from the exciting frequency is equal to the frictional losses damping such vibration.

If impellers and sleeves are clamped on a shaft, any shaft distortion causes friction to be generated between their contacts and between the impellers and the shaft such that the vibration usually is limited sufficiently to permit running through such criticals.

If this damping is insufficient, the bearings can be mounted in a damped support, such as an oil film or the like to increase the damping and permit wider operation even though there are "criticals" present.

Simple calculations or mental estimates of frequency of simple resonating systems may be made as follows:

A vibration, in general, involves a mass supported on an elastic support "spring." The resonance frequency is related to the magnitude of the mass and the stiffness of the spring by the following equation:

$$\text{Cycles per minute} = \frac{60}{2\pi}\sqrt{\frac{S}{M/g}} \tag{1}$$

where S = spring stiffness, lb/ft
                 (no. of lbs. to displace 1-ft.)
          M = lbs of oscillating mass.
          g = 32.17 (this is to convert lbs.
                 mass to slugs.)

If a term for friction is introduced into the equation such as that for friction resulting from a flexibly supported bearing having a viscous damper where the damping force is linerally proportional to the displacement velocity, instead of facing an infinitely large resonant amplitude, it becomes limited thereby.

The maximum amplitude is $\frac{X}{2} \cdot \frac{C_c}{C}$ where X is the amplitude of the normal runout, C is the degree of damping and C_c is critical damping. To visualize critical damping, envision a resonant system initially excited. If it has a moderate degree of damping the oscillation will gradually die down. With greater degree of damping it will pass only slightly through zero position on its return. Critical damping is that degree of damping which will just prevent its passing through zero on its return.

Such studies of shaft vibrations generally include an assumption that the base or frame of the machine to which the "spring" is attached is entirely rigid. This may not be so, and the mass of the frame may be comparable with that of the rotor. In such situations the mode of vibration becomes complicated and usually its frequency is reduced as a result of this.

Torsional Vibrations of Rotating Systems

The simplest torsional vibrating system would be two rotors of equal moment of inertia mounted on opposite ends of a shaft, the shaft having significant torsional flexibility (T).

T is the number of pound-feet of torque necessary to twist the shaft one radian.

$$I = (lb) (ft)^2$$
$$T = (lb) (ft)/(radian)$$

For $I_2 = I_2$, the torsional frequency is

$$= \frac{60}{2\pi}\sqrt{\frac{2T}{I/g}} \quad \text{cycles per minute} \tag{2}$$

More complicated systems than this exist such as more than two moments of inertia mounted on the shaft; the shaft having significant moment of inertia; there being a gear in this system; or there being flexible shaft coupling(s) in the system.

The designer must carefully study the torsional vibration problem as far as the machine design is concerned, but the user may become involved should his shaft coupling or drive motor influence the resonance natural frequency.

Torsional vibration isn't very common because it is usually resonant at other than operating speed, simply by probability.

In assigning shaft stiffness the following equation is applicable:

$$\frac{\text{ft-lb torque}}{\text{radian}} = \frac{GJ}{\ell}$$

where G = Shearing modulus, lb/in^2
         = 12,000,000 for steel.
J = polar moment of inertia
         = $\frac{\pi d^4}{32}$, in^4.
ℓ = length, ft.

One variable in the shaft equation, above, is its length. If the coupling hub is shrunk on, the shaft may be measured from the face of the hub. Otherwise, the effective length extends to about the center of the hub. Likewise, at the motor armature end, there is no assurance that the shaft is snugly clamped throughout, so at least a diameter of the shaft into the armature should be allowed to correct for this.

The coupling manufacturer usually will state the torsional rigidity of its couplings for use in calculations. Otherwise, they may be assumed to have about the same stiffness as an equal length of shaft.

The frequencies likely to cause problems are the running speed and twice the running speed; the latter resulting from misaligned couplings or from the electric current frequency. Three times running speed may result from the three-phase current in the motor.

When a torsional vibration amplitude gradually builds up, usually there is a keyed shaft which will first slip under the torsional oscillation. This usually darkens the slipping surfaces and pounds out the key-seats, although other fatique problems can occur.

One of the commoner torsional modes involves the low-speed shaft as the "spring"; the two inertial moments being the motor armature on one end and the gear on the other. The high-speed pinion gear system moment of inertia multiplied by the square of the gear speed ratio must be added to that of the low-speed gear.

The other common mode of vibration is in the high-speed end with the shafts and coupling as the "spring." The moment of inertia on one end is the centrifugal compressor impeller. On the other end is the gear pinion augmented by the moment of inertia of the large gear and motor system divided by the square of the gear ratio.

If this mode should occur, it could be that the coupling is unduly torsionally flexible, in which case the correction of that would solve the problem, but otherwise the diameter of the shaft or its length must be substantially changed.

If the low-speed (generator/gear) system is torsionally resonant it could be unduly influenced by the coupling. It is often presumed that a type of coupling with a zigzag spring between two notched hubs would easily solve the problem by contributing substantial torsional softness. However, this may not be the case. Such couplings are rather stiff torsionally, so that this cannot be relied upon blindly to solve the problem.

Structural Vibrations

Exciting frequencies at compressor operating speed and at the driver operating speed are usually present. If attached to or near to a machine where there are motors operating at other speeds, there may be other frequencies applied to the structure, such as half-synchronous.

It is not uncommon for the motor mounted on its base to be tuned in a lateral mode, or vertical mode, or a torsional mode to a resonant frequency near 900, 1200, 1800, or 3600 cpm (or 1000 or 1500 or 3000 cpm). If it is, it may build up substantial amplitude, such as several thousandths inch. Such resonance also may develop in the piping, the amplitude of which may reach even a sixteenth inch amplitude in some extreme cases.

At the higher frequencies, such as the compressor speed, more rigid sections of piping or other members may be excited.

The manufacturer ordinarily does not take much responsibility for this because it doesn't occur very often, and because when the unit is grouted-in and piped-up, the natural frequency may change out of the running speed, or may change into the running speed. If the latter should happen, the zone of maximum vibration should be found and identified as the "mass," and whatever supports it should be identified as its "spring." Then either the spring must be substantially changed or the mass substantially changed. This usually is somewhat harder than the serviceman first assumes. For example, if the structure identified as the spring is to be weakened by being cut through three-fourths of the way, it then might still be about 98% as "stiff" as ever. Also, in the addition of mass to the "weight" that is oscillating, the bonding of the added mass must be strong enough that insignificant flexibility exists between them.

Resonance Damage to Impellers

An impeller (tightly mounted or not) on its shaft can be excited to vibrate or resonate at several—often many—frequencies, the most common of which is the four-node peripheral axial vibration shown in Figure 1. Two opposite segments of the periphery bend in one direction while the other two bend in the opposite direction. This mode of vibration usually has a natural frequency on the order of five or ten times the maximum possible speed of the impeller. Thus the speed of rotation does not excite it, but it may be excited by some sequence of multiple pulses such as might strike the impeller from inlet vanes or from a non-uniform inlet stream striking the succession of impeller blades at the "passing frequency."

Such frequency is usually of the order of 1000 to 2000 cycles per second for an impeller a foot in diameter, and for similar impellers of different diameters, the natural frequency varies inversely with the diameter. Thus, it is the tip speed and the number of blades "the passing frequency" that excites it and it usually corresponds to a dozen blades at 500 or 1000 feet per second tip speed.

This usually is in the upper range of speed and therefore not ordinarily encountered except in high speed or heavy duty compressors.

The harmonic of the passing frequency occasionally must be considered, and this reduces the speed where resonance may occur to half.

If there is no shroud or cover on the impeller such "open" impellers are subject to blade resonance. The blades extend in cantilever fashion and will resonate in a similar range of frequencies. The blades may have overtones or different modes of vibration subject to resonance at several different frequencies.

In addition to the four-node vibration mentioned above, there may be six nodes, or eight nodes, or ten nodes, or more at higher, and higher frequencies. The ratio of the frequency of the six node to that of the four node is usually about 1.25.

There are other modes of vibration. For example, a closed impeller—one

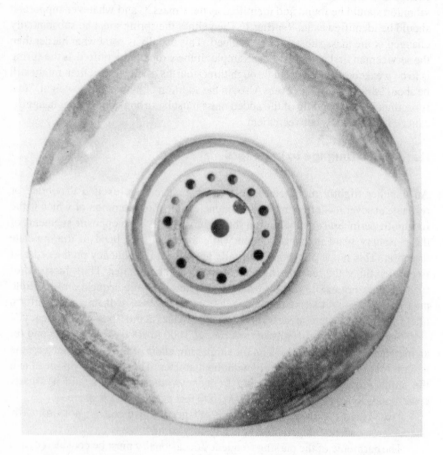

FIG. 1. Four-node peripheral resonance. Back surface of impeller in horizontal
position dusted with white sand. Upon being subjected to four-node
peripheral resonance frequency the sand bounced off of the four peripheral
vibrating segments.

with a shroud — is subject to a rotational or torsional resonance where the shroud moves in one angular direction while the disk moves in the other direction, and the shear action on the blades represents the "spring." This seems difficult to visualize, but it is a very common form of resonance and occurs at about the same range of frequencies at the others.

There are also several resonance modes to which the shroud is subject. The first one usually is the only troublesome one. It is the first order vibration of the shroud section spanning the space between the blades at the periphery of the rotor.

An impeller resonance is subject to very low loss. Therefore, in a very narrow band it is excited to a high amplitude sufficient to fatigue and break-up the wheel in a few seconds or minutes, or at least in a few hours.

Since such resonance is destructive, it is imperative that the designer measure these frequencies and determine that there are no passing frequencies in operating speed ranges that will excite any of them. It is all right to pass through them, starting and stopping. They are not perceptible, but if there were any delay going through them such as operating several minutes in any particular one of them it would probably cause some damage.

The measurement of the natural frequencies is easily done by mounting the rotor in position to be excited and exciting it with an electromagnetic shaker. The shaker can be a simple device just touching the impeller almost anywhere. The vibrating frequency (and amplitude) may be measured by having a vibration transducer touch a part of the impeller expected to resonate. It is very easily done by comparing this transducer output on an oscilloscope with a similar frequency out of phase with the transducer output such as a small voltage from the exciting frequency. In the resonance range where both frequencies are readable, a figure, ellipse or circle, called a Lissajous figure (1) shows on the screen.

To illustrate, if an impeller is found to have resonances as follows and the design speed is 12,500 rpm, then the number of blades (to cause passing frequency) to excite the impeller at each of the frequencies is listed in an adjoining column.

Natural Frequency, cps	No. of Blades	Natural Frequency, cps	No. of Blades
2870	13.8	4200	20.16
3062	14.7	4461	21.41
3206	15.39	4692	22.52
3342	16.04	4806	23.07
3632	17.43	4922	34.64
3734	17.92	5160	24.77

It is apparent that the maximum permissible number of blades is 12. It would be possible to use 13, except that a replacement impeller may vary enough to cause trouble.

It is obvious, therefore, that if any of the numbers of blades in the blade column were used, there would be damage to the impeller.

Modulation of the impeller discharge stream is usually of little consequence because the gas there is discharging, so interference with such stream would still not react on the impeller. It is the stream entering the impeller that applies momentum to it and can carry a sequence of pulses which, if at the natural frequency, will excite resonance.

Summarizing, it is imperative that the resonant frequencies of the rotor be measured and hopefully identified and that the designer give attention to avoiding any entering pulse sequence such as the blade passing frequency at the inlet or passing frequency of stream irregularities coming from inlet guide vanes or disturbed by heads of bolts or the like.

The damage usually is the notching, cracking or breaking of the shroud or blade, or the loss of a portion of a blade, any of which would cause serious damage to the compressor.

Gear Speed Increaser

Most centrifugal compressors operate in the range of 5000 to 25,000 rpm, and frequently require a gear speed increaser. The gear is subject to numerous possible problems, but the manufacturer's quality control usually protects the customer from these, such as the shape and uniformity of spacing of the teeth, and uniformity of tooth-loading, dependent on the two gear shafts being parallel.

The user may have problems with a) coupling alignment, b) coupling weight, c) pinion shaft thrust load, d) pinion bearings, e) torsional vibration, or f) possible structural vibration.

High-Speed Shaft Coupling

A popular form of shaft coupling is non-lubricated, and usually has a spool-piece between two flexible sections, such that the spool-piece may be removed for easier access to the compressor. It has the valuable quality of requiring substantially no maintenance because no lubrication is required. Such couplings usually will tolerate an angle up to a quarter of a degree on either end. With misalignment in excess of that, the flexible members may become fatigued and fail. A possible problem with these couplings in the high-speed models is the centrifugal force on bolt heads and nuts if they extend outward excessively such as when spacers are placed under the nut or head. They also stiffen axially several-fold at speed.

Lubricated Couplings

High-speed couplings that are lubricated may require only filling with oil or grease. In such case, success depends on the grease not leaking out between service periods. If the coupling is continuously lubricated, as they often are, the oil must be filtered to avoid the coupling accumulating a deposit of particles centrifuged out of the oil, and this requires filtering to 5 microns or better.

The gear often is a double-row helical gear, such that the thrust due to the helical teeth balance each other. If there should be a superimposed shaft thrust load it will transfer more of the load to one of the rows of teeth and can easily overload it. This must be given careful attention, knowing also that in the case of non-lubricated couplings, they stiffen axially usually several-fold at speed. With gear-type couplings, the gear load may develop an end thrust, but usually there is enough misalignment or vibration in the system that they will seek a low-thrust position.

The low-speed coupling may be of more conventional design, such as grease-packed. Sometimes they should be locked axially to carry the motor near its mid-position, relying on the bullgear thrust bearings to locate it.

See also shaft torsionals further above.

Shaft Seals for Compressors

The shaft seals generally available for the shafts of compressors are:
     a. fixed clearance (labyrinth type seals),
     b. floating rings, usually carbon ring seals (like piston rings),
     c. "mechanical" seals,
     d. lubricated seals such as lubricated journal bearings acting as positive seals,
     e. combinations of the above,
     f. special floating seals and the like.

Mechanical Seals:

These seals are mechanically reasonably simple, are widely available and they are positive seals, both while running and while shut down. They would often be first choice unless there are some objectionable problems.

Many factors affect their performance and life. However, with reasonable conditions they can be operated up to several hundred psi and even higher at reasonably high speeds. Their limitations are published in various places (2), (3).

On the negative side, they have limited life usually of the order of a year; they require cooling, either by the pump liquid or by circulating lubricant; they take more than a negligible amount of axial space on the shaft which at high speed is often at a premium.

They are frequently installed in pairs with the pressurized lubricant between the two seals at a pressure above that to be sealed (these are described in detail in this volume, p. 180).

Floating Carbon Rings:

This is not a positive seal, but it usually seals somewhat better than a fixed clearance seal. It has the objection that the shaft usually is worn by the rings rubbing on it.

They can be put in pairs in the event that a buffer gas is to be injected mid-seal.

Fixed Clearance Seals:

These ordinarily comprise a rotating journal with a "labyrinth-type seal" in close proximity to it, of the order of a few thousandths inch clearance. The labyrinth type is a series of knife edges usually about 5/1000 inch wide at the edge.

These knife edges are usually several or more in number. They can be put in groups with a "lantern ring" or annulus between the groups into which (or out of which) a stream of buffer fluid or the like may be injected or withdrawn. A common spacing is 1/16 to 3/16 inch.

The materials need to be reasonably compatible so that they do not rub and "pick up" or seize unduly. Stainless steel is a difficult material; nevertheless, hardened stainless steel shafts engaging thin stainless steel edges are reasonably successful.

At higher pressures the fixed clearance seals (labyrinth-type seals) leak at rapid mass flow rate because the mass flow is proportional to the density of the gas, approximately proportional to gas pressure. Accordingly, at elevated pressures, it is often the practice to control the seal leakage by means of a positive seal of some kind and an orifice to restrict the seal leakage flow rate. This will reduce the leakage to about half that which it would be if leaking freely to atmospheric pressure. Instrumentation can further reduce the leakage such as with the use of differential pressure instruments or, in the case of different temperatures across the seal, by the use of a thermocouple to sense the direction of flow (3).

There is also an array of seals for low leakage comprising fixed clearance surfaces maintained by gas bearing type thrust bearings. These are usually specialty seals where the characteristics of a gas bearing is permissible or tolerable (4), (5), (6).

Optimization

In the final design of a centrifugal compressor stage, the effects of all variables on performance (including operating environment) and cost should be determined or estimated, and the final design established by maximizing the performance within the cost constraint.

This is a difficult task, complicated by need for data, expensive to obtain. For these and other reasons many, if not most, designs have room for further improvement. The goal is approached asymptotically to the extent that talent and R & D money are put into it.

References

1. H.V. Malmstadt, C.G. Enke, and E.C. Toren, Jr., *Electronics for Scientists*, W.A. Benjamin Inc., New York, 1963, pp. 30–34.
2. Wm. J. Mead, *The Encyclopedia of Chemical Process Equipment*, Reinhold Publishing Corp., New York, 1964, pp. 825–831.
3. U.S.Patent No. 3,375,015 (1968).
4. H.S. Cheng, et al., "Performance Characteristics of Sprial-Groove and Shrouded Raleigh Step Profiles for High-Speed Noncontacting Gas Seals," Paper No. 68-Lubs-20 (1968).
5. Sealol, Inc. "Technical Bulletin" No. 1-67 (March 1967).
6. U.S.Patent No. 2,673,752 (1954).

COMPRESSORS
AND EXPANDERS

General Type Selection Factors

HEINZ P. BLOCH

Compressors are used in petrochemical plants to raise the static pressure of air and process gases to levels required to overcome pipe friction, effect a certain reaction at the point of final delivery, or to impart desired thermodynamic properties to the medium compressed. These compressors come in a variety of sizes, types, and models, each of which fulfills a given need and is likely to represent the optimum configuration for a given set of requirements.

Compressor type selection must, therefore, be preceded by a comparison between service requirements and compressor capabilities. This initial comparison will generally lead to a review of the economies of space, installed cost, operating cost, and maintenance requirements of competing types. Where the superiority of one compressor type or model over a competing offer is not obvious, a more detailed analysis may be justified.

1

Compressor Types

Compression machinery can be separated into two broad categories: dynamic and positive displacement. The centrifugal compressor is a dynamic machine by contrast to the static, positive-displacement type of compressor. Centrifugal compressors do their work by using inertial forces applied to the gas by means of rotating bladed impellers, whereas positive-displacement compressors trap gas by the action of mechanical components and restrict its escape as compression takes place through direct volume reduction. Each of these two broad categories can be further subdivided as shown in Fig. 1.

A number of service conditions affect the type selection. Foremost among these are volume flow rate and discharge pressure. Both of these parameters, in turn, influence power levels.

Generally feasible flow and discharge pressure ranges are given in Fig. 2. Note, however, that these ranges are given for guidance only. Compressor design is a fast-moving field and state-of-arts advances may shift this picture within relatively short time periods.

Limitations of Various Compressor Types

Aside from the limitations of flow and discharge pressure evident from Fig. 2, compressor type selection will be affected by the characteristics of the process system. Table 1 lists these suitability criteria.

Quite evidently, each of the different types of compressors has characteristics which make it more suitable for some application than for another. In order to assist in eliminating certain compressors from study early in the design, some of these features are listed below.

Reciprocating Compressors

Reciprocating compressors are available for almost all compressor appli-

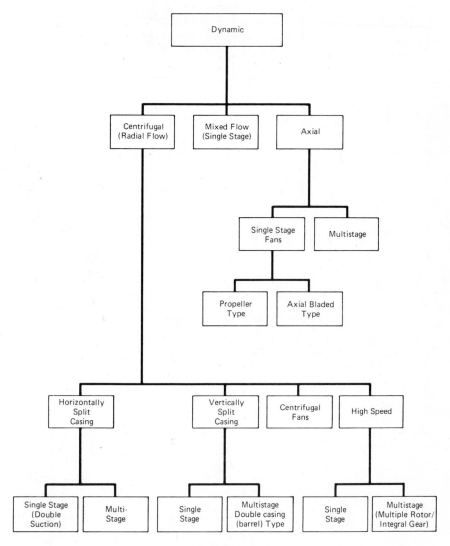

FIG. 1.   Chart of principal compressor types.

TABLE 1 Limitations Imposed on Certain Compressor Types by Characteristics of Process System

Characteristic of System	Suitability of Given Compressor Types			
	Generally Suitable	Conditionally Suitable	Requires Expensive Added Investment	Usually Unsuitable
Liquid carry-over into compressor is likely	Liquid ring compressor Helical screw compressor	Standard centrifugal, axial	Reciprocating, sliding vane, high-speed centrifugal	
Solids carry-over into compressor is likely	Liquid ring compressor	Helical screw, standard centrifugals, axials	Reciprocating, sliding vane, high-speed centrifugal	
Does not tolerate lube oil intrusion	Nonlubricated reciprocator	Standard centrifugals, axials, liquid ring compressor	Helical screw	Lubricated reciprocating, sliding vane compressor
Molecular weight of gas varies	Reciprocating, sliding vane	Helical screw, liquid ring compressor	Centrifugals, axials	
High inlet temperature likely		Centrifugals, axials, helical, screw compressor		

	First Choice	Second Choice	Third Choice	Fourth Choice
Fouling tendency of gas	Helical screw compressor	Conventional centrifugals		Lubricated reciprocating compressor, high-speed centrifugal compressor, axial compressor

Special Requirement	First Choice	Second Choice	Third Choice	Fourth Choice
High efficiency is required	Reciprocating sliding vane	Axial compressor, helical screw compressor	Centrifugal	Liquid ring compressor
Low maintenance costs required	Centrifugals, axials	Helical screw, liquid ring	Reciprocating (lubricated)	Sliding vane, reciprocating compressor (nonlubricated)
Very low flow	Low flow–high head centrifugal	Diaphragm compressor		

cations. They are suitable for all pressures from vacuum to around 100,000 lb/in.^{2}gauge. They are available to handle volumes from less than 5 ACFM up to 7,000 ACFM. Their overall efficiency varies from 80 to 90%, averaging about 85%.

The disadvantages of these machines are:

For continuous duties such as powerformer hydrogen recycle compression, more than one machine must be provided to permit servicing.
They are large and expensive.
They have a high maintenance cost, especially when handling gases containing liquids, solids, or corrosive materials.
Because they have large shaking forces, large foundations are required.

Rotary Screw Compressors

This compressor type is generally available for pressures up to 250 lb/in.^{2}gauge and for volumes of 800 to 20,000 ACFM. They are balanced machines and require only light foundations. Because there are no rubbing surfaces, they do not contaminate the compressed gas with lubricating oil. The maintenance cost should be low. In their operating range they are cheaper than both centrifugal and reciprocating machines. Their efficiency is in the range of 75 to 85%. They can be used to compress gases containing tars or polymers, and in these cases the efficiency of the machine is higher than when handling clean gases.

The disadvantages of these machines are:

The machines are noisy.
The range of capacity variation possible at constant speed is very small.
The machines are designed for a specified gas and compression ratio.

Sliding Vane Compressors

Sliding vane compressors are available for pressures up to 150 lb/in.^{2}gauge. They are suitable for volume ranges of 50 to 6,000 ACFM. They do not require foundations as heavy as reciprocating machines. If the operating conditions are suitable, they are cheaper than reciprocating machines and they require less maintenance.

The disadvantages of these machines are:

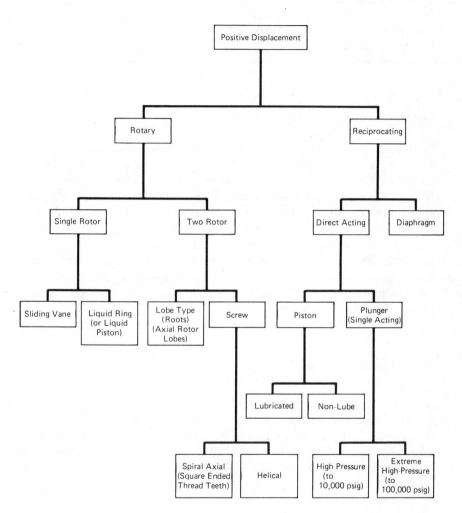

FIG. 2. Approximate ranges of application for reciprocating, centrifugal, and axial flow compressors.

In pressure service they are limited in compression ratio to about 3.5/1 and to a
    differential pressure of about 60 lb/in.^{2} in each casing.
The compressed gas leaving the machine contains excessive quantities of
    lubricating oil.
They should only be used to handle clean gases.
The efficiency will vary from about 60 to 75%.
Multiple installations are required to permit servicing unless the duty is
    intermittent.

Centrifugal Compressors

Centrifugal compressors are extensively used in modern petrochemical plants.
They are basically large volume machines. They are available for pressures of
up to over 5,000 lb/in.^{2} gauge and handle volumes of 1,000 to 150,000 ACFM.
Because there are no rubbing surfaces, they do not contaminate the compressed
gas with lubricating oil. They are balanced machines and do not require heavy
foundations. Their efficiency is in the range of 68 to 76%. The maintenance cost
is low. In their operating range, their initial cost is less than that of reciprocating
machines. The capacity can be controlled by speed variations, reducing the
suction pressure, or by inlet vane control. The service factor is so high that only
one compressor is required even in services requiring 3 or more years in
continuous operation.

Axial Flow Compressors

Axial flow compressors are essentially very high volume machines. Except at
volume flows over 60,000 ACFM, they are more costly than centrifugal
compressors. The efficiency of an axial compressor is as much as 10% higher
than it is for a centrifugal compressor. Discharge pressures of 200 lb/in.^{2} gauge
have been demonstrated on axial flow process machines.

Thermal Compressors

Thermal compressors, also known as ejectors, are venturi tubes with a gas
nozzle in the eye, usually using steam as the driving fluid. They are quite
inefficient and normal usage is in vacuum service where small quantities of gas

are handled. They can, by multistaging, draw pressures down to 1 in. of mercury absolute. They are used on vacuum tower service and steam condensers, and small models are used to extract steam from the glands of turbines.

Special Applications

Other types of compressors are available for special service. An example is the use of rotary lobe compressors of the Roots type for very low pressures, i.e., suction, 0 lb/in.^{2}gauge; discharge, 10 lb/in.^{2}gauge. Another example would be the use of water-sealed positive displacement rotary compressors in vacuum service. These applications are unusual and are not dealt with in this text.

Limiting Parameters

Horsepower

Reciprocating machines are available up to 15,000 hp per machine. Sliding vane compressors are available up to 400 hp per machine. Rotary screw compressors are available from 10 to 6,000 hp per machine. Centrifugal machines are used on light molecular weight gases above 2,000 hp and on general hydrocarbon gases above 500 hp. Single wheel centrifugal machines are available at lower horsepowers for low head applications, such as blowers, but Sundyne has a machine up to 400 hp and up to 25,000 ft of head.

Volume

A centrifugal machine will probably be uneconomic if the suction volume is below 2,000 ACFM and the discharge volume is below 500 ACFM. Maximum volume is 150,000 ACFM. A rotary screw Lysholm-type compressor will probably be uneconomic if the suction volume to any casing of the machine is less than 1,000 ACFM or more than 20,000 ACFM. Sliding vane compressors may be economic within the range of 50 to 3,000 ACFM. Reciprocating compressors may be used for all capacities up to about 7,000 ACFM.

Pressure

Centrifugal compressors can be made for discharge pressures of over 5,000 lb/in.^{2}gauge. The maximum adiabatic head per casing is about 100,000 ft.

Rotary screw compressors have a maximum discharge pressure of 250 lb/in.^{2}gauge. The maximum compression ratio per casing is about 4.5/1. Sliding vane compressors have a maximum discharge pressure of 150 lb/in.^{2}. The maximum compression ratio per casing is about 3.5, and the maximum differential pressure per case is about 60 lb/in.^{2}.

Reciprocating compressors have a maximum discharge pressure of about 100,000 lb/in.^{2}gauge. The compression ratio per cylinder depends upon the k value of the gas and the inlet temperature. Diatomic gases at 90°F suction can have a compression ratio up to 4/1, while hydrocarbon gases with a k of 1.2 can have a compression ratio of up to 9/1. With high suction pressures, blowby on the piston or piston rod loadings may limit the compression ratio.

Gas Conditions

It is undesirable to have liquids or solids in the gas stream to any compressor. Centrifugal compressors can handle liquids and solids better than other types. Tar and polymer contaminants are best handled by Lysholm-type compressors. Sliding vane compressors can handle some liquid, but the quantity should not be sufficient to dilute the lubricant. Axial compressors are not suitable for fouling gas services since the deposits will greatly reduce their efficiency.

Reciprocating and sliding vane compressors are very susceptible to corrosion and wear unless the jacket temperature is maintained above 90°F. This prevents condensation in the cylinder and maintains lubricity of the oil.

Weather Protection

All the types of compressors listed earlier are suitable, or can be made suitable, for operation in the open, without weather protection. Even a shelter such as is normally provided in typical petrochemical plants is not required for the compressor. Whether some form of protection is necessary for operating personnel and maintenance crews is debatable. In any event, it is unlikely that more protection needs to be provided than is required for pumps.

Centrifugal and Axial Turbocompressors*

HEINZ P. BLOCH

The rapid growth of the chemical industry in recent years has resulted in plants of ever-greater capacity and a large number of new processes, thus opening up a wide range of applications for turbocompressors. The largest and fastest growing user of these machines is the vast industry centering around petrochemicals derived from oil and natural gas.

Large numbers of centrifugal compressors (Fig. 1) and axial compressors (Fig. 2) are used. Today there are at least 120 main chemical processes in which turbocompressors are employed for transporting gaseous fluids or compressing them to the pressure levels required by chemical reactions.

The properties of the gases handled, the flow rates, the pressure difference to be overcome, and the operating temperatures can vary within very wide limits, depending on the type of service and production capacity. To satisfy these ever-changing operating conditions requires a broad range of turbocompressors in different sizes and designs. Thus design flexibility must be engineered into these standard sizes.

*Based on A. Buchel, *Sulzer Turbocompressors in the Chemical Industry*, Sulzer Bros., Zurich, Switzerland, 1974.

FIG. 1. Typical centrifugal compressor, horizontally split casing arrangement. (Courtesy Sulzer Brothers.)

Added to this, the chemical industry imposes some very exacting design requirements which are not necessarily required in other applications. For example:

Special care in selecting materials where corrosive gases, high pressures, or very low temperatures are involved
Gastight shaft seals are needed in the majority of cases
Limitation of temperature during compression in order to avoid polymerization
Injection of solvents to clean the flow paths in the case of dirty gases
Extreme care is necessary in balancing the rotor, designing the coupling, and overcoming axial thrust in the case of small compressors operating at high pressures

Theoretical Considerations

The essential design parameters of a turbocompressor are:

The reference diameter D, which is related to the frame size and corresponds to the impeller tip diameter of a centrifugal compressor or the drum diameter of an axial rotor
The number of compression stages z

The equations governing the thermodynamic layout of a compressor to suit particular operating conditions can be summarized as follows:

$$D = \sqrt{\frac{V_1}{\phi U}} \quad (m) \tag{1}$$

$$z = \frac{RT_1 \dfrac{n}{n-1}\left[\left(\dfrac{p_2}{p_1}\right)^{(n-1)/n} - 1\right]}{\mu U^2} \tag{2}$$

$$N = \frac{60U}{\pi D} \quad (\text{rev}/\text{min}) \tag{3}$$

FIG. 2. Typical axial compressor with adjustable stator blades. (Courtesy Sulzer Brothers.)

where
V_1 = suction volume (m³/s)
U = peripheral speed at reference diameter (m/s)
R = gas constant = 8315/molecular weight (J/kg °K)
T_1 = temperature at suction conditions (°K)
P_2/P_1 = pressure ratio
n = polytropic coefficient
ϕ = flow coefficient
μ = pressure coefficient
N = rotational speed (rev/min)

In Figs. 3 and 4 the required diameter and number of stages are shown as a function of the other parameters, in accordance with Eqs. (1) and (2). The compressor design, i.e., the selection of diameter D and number of stages z, is subject to the following constraints.

Peripheral speed U: As can be seen from Eqs. (1) and (2), higher peripheral speeds U permit smaller machines. This speed, however, is limited by the strength of the rotor material or the Mach number, whichever is lower. Mach number is defined as the ratio of peripheral speed U to the acoustic velocity $(kRT_1)^{1/2}$ in the gas under the conditions prevailing at the impeller inlet. With industrial subsonic compressor stages this number should never exceed 0.95 as otherwise hydraulic losses increase sharply. Figure 5 indicates the maximum permissible peripheral speed.

Flow coefficient ϕ: This coefficient depends on the geometry of the flow path in the compressor and is determined by the relative impeller width b/D in centrifugals or relative blade height h/D in axials, as well as by the vane or blade shape. It can be selected within certain limits beyond which the polytropic efficiency drops to unacceptably low values (Fig. 6).

Pressure coefficient: This depends on the geometry of the impeller vanes or the setting angle of the axial blades, and determines both the stage efficiency and the compressor turndown capability. To keep the latter sufficiently wide, the value of this coefficient is limited to about 0.48 for centrifugals and 0.35 for axials.

Critical speed: Modern turbomachines normally operate at between the first and second natural bending frequencies of their rotors. Increasing the length of a rotor so as to accommodate more stages brings the second critical speed, which varies in inverse proportion to this length, closer to the operating speed. According to API standards, a margin of at least 20% should be observed, thus limiting the maximum rotor length and hence the maximum number of stages. This limit is about eight impellers for a centrifugal machine, but may be increased by up to 50% if narrow impellers are used for special applications. With axials the number of stages can be as many as 24, but is much

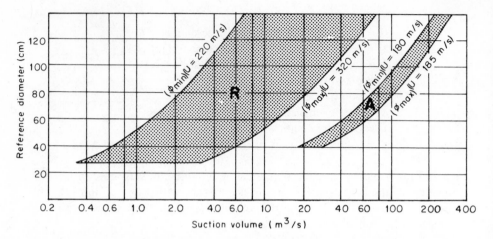

FIG. 3. Compressor reference diameter in relation to the suction volume, the peripheral speed U (m/s), and the flow coefficient ϕ (Eq. 1). Area R is for centrifugal compressors and Area A is for axial compressors.

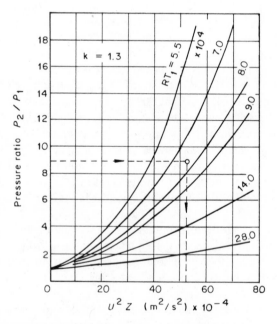

FIG. 4. Number of centrifugal compressor stages z as a function of the pressure ratio P_2/P_1 and other parameters (Eq. 2). T_1 is the suction temperature (°K), U is the peripheral speed (m/s), R is the gas constant (J/kg · °K), and $RT_1 = $ J/kg.

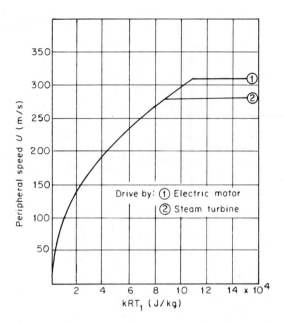

FIG. 5. Peripheral speed of a centrifugal compressor as a function of the acoustic velocity in the gas under conditions prevailing at the compressor inlet. In the case of heavy gases with inherently low acoustic velocity, the peripheral speed is limited by the Mach number, whereas with light gases the limiting factor is the tensile strength. With variable speed drive, this limit has to be determined with regard to the trip speed which will be approximately 110% of the maximum continuous speed.

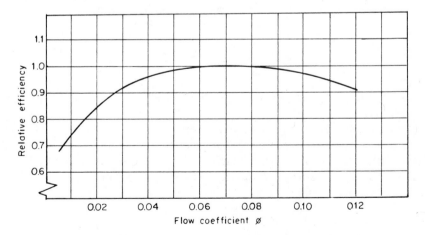

FIG. 6. Centrifugal impeller efficiency as a function of the flow coefficient. Due to the decrease of volume flow during compression between suction and discharge, the widths of the impellers become smaller. The ϕ values therefore decrease, and the stage efficiencies deteriorate. Some turbocompressors employ first stages having a high flow coefficient so that downstream stages are still relatively wide, thus ensuring high overall efficiency.

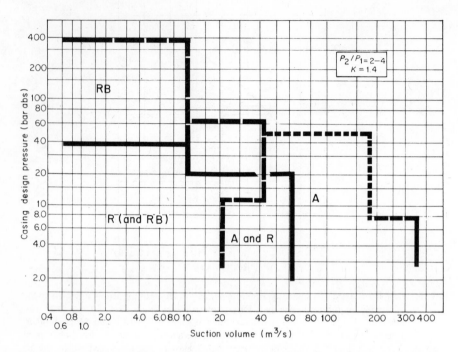

FIG. 7.   Range of compressor types (typical only) in relation to the suction volume. Each frame
size is available for different numbers of stages and different design pressures. R indicates
centrifugal compressors with horizontally split casing. RB indicates centrifugal compressors with vertically split casing (barrel). A indicates axial compressors.

less when high operating pressures dictate strengthened blades with longer
chords.

Temperature rise: High operating temperatures lead to high power
consumption, and may encourage polymerization in some kinds of gases. Also,
with centrifugal machines the diaphragms contained in the casing may undergo
excessive radial expansion if the compressed gas becomes too hot. The
temperature rise during compression is limited by providing intermediate
cooling between the groups of stages.

Compressor Design Adaptability

The different ways of adapting standard designs to varied operating conditions are described below.

Adapting to Different Flow Volumes

The diameter D, i.e., the compressor frame size, is derived from Eq. (1), bearing in mind the limitations imposed by peripheral speed U and the flow coefficient ϕ.

Type Selection—Centrifugal or Axial

Figure 7 shows a wide overlapping area in which centrifugal and axial machines compete. Above a minimum suction flow volume of 20 m^3/s depending on the delivery pressure required, axial compressors offer certain advantages. For a given pressure ratio their polytropic efficiency is between 3 and 7% better, and their dimensions are smaller than those of a corresponding centrifugal machine. For the same flow their rotor tip diameter is roughly 40% smaller than for the centrifugal counterpart, and consequently the speed of rotation in rev/min is about 40% higher. For high-output low-pressure compressors of at least 25 to 30 MW, these higher speeds are ideally suited to direct drive by means of 2-pole electric motors or steam or gas turbines of proven design. Such machines are used for driving generators at 3000 or 3600 rev/min and do not require gearing. Gears may not always be readily available for these speeds and powers.

On the debit side, the pressure ratio developed in one axial casing is smaller than that possible in a centrifugal machine. Centrifugals may be preferred for this reason.

Which type of machine to select is ultimately decided by operational and economic factors. These must be considered in their entirety before a final choice can be made.

Adapting to Different Operating Pressures

Since turbocompressors may be used as low-pressure machines or as high-pressure boosters and therefore operate at different pressure levels, their standardized casings are engineered for selected ratings. These have been chosen with the requirements of the compressor market in mind, which demands both low and high design pressures with the smaller frame sizes, whereas high flow machines usually operate only at lower pressure levels (see Fig. 7).

The casings of centrifugal compressors can be split either horizontally or vertically. Rotor maintenance is easier with horizontally split casings than when the casing is split vertically. However, a horizontally split casing is limited as regards pressure, owing to the large sealing area of the joint. The API standards widely applied in the chemical industry have recently introduced a guideline that calls for a vertical joint or barrel casing under the following conditions:

Mole fraction of H_2%	100	90	80	70
Maximum casing operating pressure (bar abs)	13.8	15.3	17.3	19.7

For heavy gases with low H_2 content, horizontally split casings are used up to operating pressures of 60 to 80 bar.

Adapting to Different Pressure Ratios

The pressure ratio and characteristics of the gas dictate the number of compressor stages. The latter is found from Eq. (2), taking into account the constraints imposed on peripheral speed U, pressure coefficient, and rotor length with respect to critical speed. The required pressure ratio is obtained by selecting the appropriate speed.

Compressor frame sizes marketed by major manufacturers exist in different standardized lengths to accommodate different numbers of stages. With cast casings, bearing span variations can be achieved by using a modular technique whereby spacing rings are fitted between standardized pattern parts of the casing to alter the length. Fabricated steel casings are feasible and are available from many compressor manufacturers.

When the head requirement exceeds the capability of one compressor body, two or more casings can be connected in series to form a train. The compressors can all be driven at the same speed if the changes in flow volume from one casing to the next are small. With heavy gases, speed-increasing gears are fitted between the casings to obtain the optimum speed for each.

Adapting to Different Process Conditions

The process industry has ever-changing requirements for location and arrangement of external casing nozzles. For instance, intercooling may be needed to limit temperature during compression, or intermediate inlets have to be provided for sidestreams. Casing designs by major manufacturers are very flexible in this regard.

Centrifugals

Centrifugals with horizontally split casings can be furnished as straight-through compressors without provision for intercooling; compressors with 1, 2, or 3 intercoolers; or compressors with 1, 2, or 3 intermediate inlet nozzles for 1, 2, or 3 sidestreams. The casings of these three series are generally similar and capable of employing the same internal parts.

Centrifugals with vertically split casings are available as straight-through compressor without provision for intercooling, compressor with 1 pair of intermediate nozzles for connection to an intercooler, or compressor with additional inlet and outlet for a recycling stage (as used in high-pressure synthesis processes).

Axials

Since the pressure ratios attained in axial compressors do not normally demand temperature limitation, these machines are designed as straight-through flow compressors without intermediate nozzles for intercooling. For high flow volumes and pressure ratios with air of 6.5 to 10, combined axial/radial compressors are available which have one intercooling stage between the sections (see Fig. 8).

Axial machines are also being designed with intermediate nozzles for sidestreams.

General Design Features of Centrifugal Compressors

Figure 9 shows cross sections through different types of machines. These consist of the following main parts.

The outer casing, split horizontally or vertically. All suction, intermediate, and discharge branches are integral with the casing cylinder and face downwards or, optionally, upwards.

Horizontally split casings are made of cast iron, nodular cast iron, cast steel, or fabricated steel according to the operating pressure and gas characteristics. For flammable or toxic gases the casing split flange may incorporate provisions for controlled leakage or inert-gas sealing. The bearing housings are generally

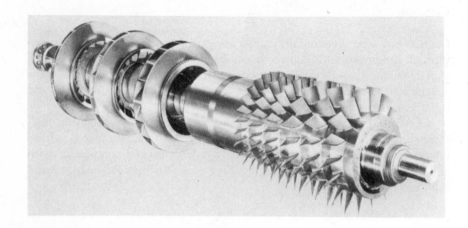

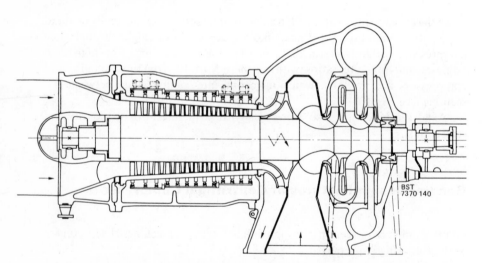

FIG. 8. Top: Rotor of an axial-centrifugal combination. Bottom: Axial-centrifugal combination with concentric piping to an intercooler. (Courtesy Sulzer Brothers.)

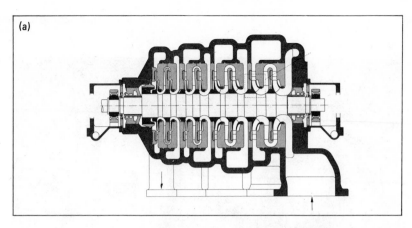

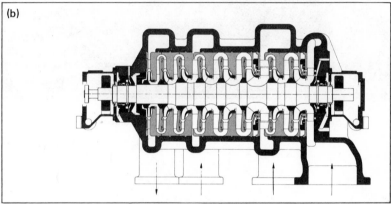

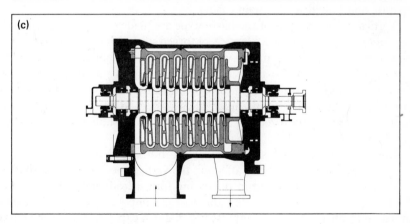

FIG. 9. Cross sections through different types of centrifugal compressors. (a) Arranged for built-in intercooling. (b) Horizontally split side-stream casing. (c) Barrel-type, vertically split casing. (Courtesy Sulzer Brothers.)

23

FIG. 10.   Barrel compressor: assembly.

separate and bolted to the main casing.

Vertically split casings of the barrel type are made of cast steel, fabricated steel, or for very high-pressure service, of forged steel with welded nozzles. They are closed by an internal autoclave cover and by one bolted end cover. The inside parts, including a two-piece internal housing, the autoclave cover, the diaphragms, and the rotor with bearings and seals, are assembled outside the barrel casing, thus allowing the internal clearances to be thoroughly checked prior to final assembly. Erection and dismantling are shown schematically in Fig. 10.

Both casing types are supported at the horizontal centerline to permit expansion in diameter without affecting the vertical position of the shaft.

Cast partition walls fitted in the casing form aerodynamically optimized guide and return ducts between the various impellers. Efficiency can be improved at the expense of compressor turndown capability by fitting curved vanes in the parallel-walled diffusors.

A shaft of solid forged steel supports the impellers. Some shaft designs embody circumferential contours near the impeller inlets to provide optimum flow approach conditions to each stage. The impellers, with backward-curved and sometimes twisted blades, are generally of welded design, although riveted designs are available in some cases.

The impeller and shaft materials undergo a number of metallurgical tests. Every impeller is balanced and overspeed tested. After assembly on the shaft, the whole rotor is dynamically balanced.

Adjustable inlet guide vanes, a standard option with most manufacturers, can be accommodated before the first stages or, in certain cases, also before intermediate stage groups. The vanes can be linked to a pressure or volume flow control system. Optimum prerotation is thus imparted to the flow at a given operating condition. This results in extended turndown range.

The vanes are pivoted in self-lubricating bushings. The angle is adjusted by moving a ring connected to the shafts through levers.

Different types of standardized journal and thrust bearings, as well as normal labyrinth seals or special shaft seals, can be accommodated. These elements are described later.

General Design Features of Axial Compressors

Originally, axial compressors were used primarily for air as blast furnace blowers. Today, they also find wide application in the chemical industry.

Cross sections through turbocompressors with fixed and adjustable stator blades are shown in Fig. 11. Both types share the same basic design elements and consist of the following parts.

The outer casing, horizontally split, is made of cast iron, nodular cast iron or cast steel, depending on the operating pressure. Suction and discharge nozzles are cast integral with the lower half, permitting removal of the upper casing half without disconnecting the large piping. The bearing housings are bolted to the casing in the smaller frame sizes, allowing the machine to be transported fully assembled. Larger frame sizes have separate bearing pedestals. The casing split flange is provided with drainage grooves for sealing purposes or to allow controlled leakage of very small quantities of gas.

The blade carrier is suspended in the outer casing. This double casing design minimizes internal stresses and distortion. It also facilitates assembly of the stator blades and routine inspection.

The adjustable stator blades (Fig. 12) are supported in self-lubricating graphite or sintered metal bearings. The latter are shrunk in steel sleeves fixed in the blade carrier. Adjusting levers are secured to the pivots and engage in grooves in the adjusting cylinder.

(a)

(b)

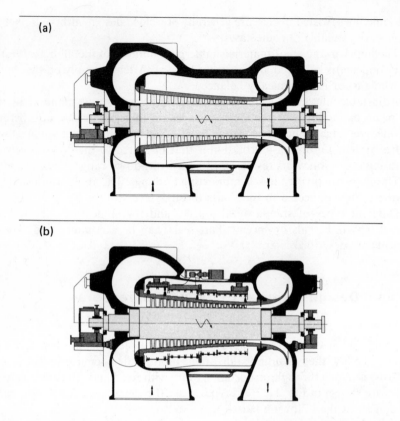

FIG. 11.   Cross sections through axial compressors with (a) fixed and (b) adjustable stator blades.

The adjusting cylinder is actuated by lateral high-pressure hydraulic servomotors and moves axially, thus varying the setting of all the stator blades simultaneously.

The rotor is a solid monoblock forging for the smaller frame sizes or of hollow welded construction for larger models, thus reducing the inertia values. Careful balancing in three planes at full speed is required to ensure trouble-free operation.

The rotor blades are held in fir-tree slots, keeping the blades firmly and accurately in position when the compressor is running. They are generally machined from solid blanks. Blade dimensions must be selected to remove their natural frequencies from harmful excitation due to periodic aerodynamic forces.

As with centrifugals, axial compressors can be fitted with any type of bearings and shaft seals as described in the following paragraphs.

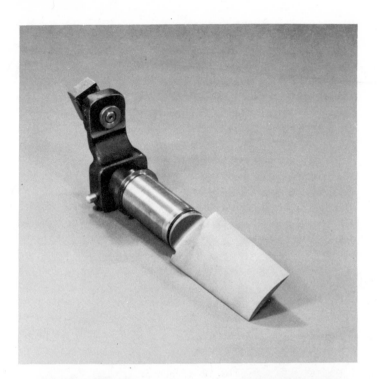

FIG. 12.   Rotor blade and adjustable stator blade of an axial compressor. The stator blade rotates in a self-lubricating, completely maintenance-free bearing. (Courtesy Sulzer Brothers.)

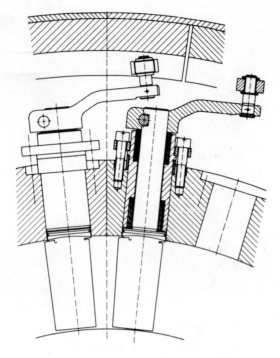

FIG. 12. Continued

Shaft Seals and Sealing Systems

Most standard centrifugal and axial compressors can be equipped with different seals to effectively prevent the passage of gas between rotating and stationary parts at the shaft penetrations.

Labyrinth Seals (Fig. 13a)

For low pressures and applications where a constant loss of the compressed gas to atmosphere can be tolerated, the simplest seal is the well-known labyrinth type. API-type labyrinths consist of "teeth" machined into a stationary ring surrounding the shaft. Some European manufacturers prefer thin steel or special alloy strips, caulked into grooves of the rotor. Either type of labyrinth is replaceable.

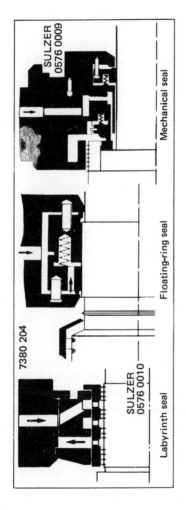

FIG. 13. Shaft seals for typical turbocompressors: (a) Labyrinth seals, (b) floating ring seals, and (c) mechanical seals. (Courtesy Sulzer Brothers.)

Labyrinth seals may be subdivided in sections so that a buffer gas or a venting system can be used in low-pressure applications where dilution with a buffer gas is permitted.

Floating Ring Seals (Fig. 13b)

Gas-tight shaft seals of the noncontact type are the basic recommendation for compressors delivering flammable, toxic, or expensive gases. They consist of two pairs of floating rings located on both shaft ends and fitted so that they are free to follow movements of the shaft relative to the casing. Sealing liquid, generally oil, is fed into the space between the two rings and maintained at an inlet pressure higher than the pressure of the gas. A small amount of sealing oil escapes through the annular gap between the shaft and the ring on the atmospheric side where it is intercepted by labyrinth strips. Similarly, a film of oil flows out through the clearance of the ring facing the sealed gas space, preventing gas leakage.

This leakage of sealing oil toward the gas side is acceptable in most cases, since the oil can easily be recovered after suitable degasing.

Well-designed floating rings have gas leakage rates of less than 20 cm³/min. Floating rings operate satisfactorily at peripheral speeds as high as 100 m/s and are therefore preferably fitted on high-speed machines or compressors with large shaft diameters.

Mechanical Seals (Fig. 13c)

Most turbocompressors can also be fitted with mechanical shaft seals. In addition to the two floating rings as described above, a sliding contact ring supported in the casing is provided on the gas side. This is pressed against the front surface of a shaft sleeve collar by peripheral springs and by the sealing oil pressure. This kind of seal can be used for velocities up to about 65 m/s and has gained wide acceptance in centrifugal compressors where sliding velocities are relatively low. However, it should primarly be used for applications where the contact rings can be replaced periodically, although they have been installed very successfully in cases where minimum intervals between overhauls are in excess of 10,000 operating hours.

Stand-Still Seals

When compressors have to stand idle for relatively long periods with the sealing system shut down, gas leakage may be prevented by means of stand-still seals. A spring-loaded sealing piston incorporated in the shaft seal closes when the machine is shut down.

Sealing Systems (Fig. 14)

In sealing systems built in accordance with API specifications, oil delivered to the liquid-film shaft seals is supplied from a reservoir by two seal-oil pumps via twin filters and a cooler. A regulation system keeps the pressure in the oil system higher than the gas pressure ahead of the seals by a constant margin.

For high pressures (above 100 bar) or when oil leakage flow toward the process gas is to be minimized, the differential pressure between seal oil and gas must be kept very small. The preferred control method is then achieved by means of the static head of an oil tank located higher than the compressor shaft. This gravity control method allows much greater accuracy in setting a small differential pressure than does a conventional pressure controller actuating a valve.

For medium pressures, below 100 bar, and especially if the above restrictions do not apply, the differential pressure of about 1 to 2 bar is adjusted by a pressure controller. The elevated tank with check valve is maintained full of oil and is placed at a level much lower than needed for the differential-pressure system described above.

With both systems the entry of seal oil into the compressor is prevented by means of a system which allows a minimum gas flow through the gas side-ring. This gas can be recovered (dried and purified if necessary) or flared.

The small quantity of oil drained off from the gas side is recovered in separate, level-controlled pots, where pressure is reduced to atmospheric. It is fed back to the reservoir through vapor traps and degassing tanks.

Rotor Dynamics and the Effect of Journal Bearings

Turbocompressors normally operate between the first and the second critical speeds of their rotor. Their dynamic behavior depends on such parameters as the rotor shape, the masses the rotor is supporting, and the elasticity of the lubricant film in the bearing.

Various criteria for admissible shaft vibrations have been computed by different standards institutions. It should be pointed out, however, that it is not

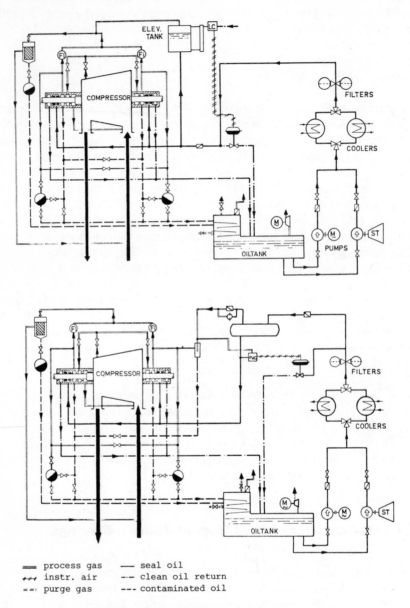

FIG. 14. Schematic flow sheets of seal oil systems. Top: Elevated seal oil tank with level control. Bottom: Differential pressure control.

FIG. 15. Tilting-pad journal bearings. (Courtesy Sulzer Brothers.)

sufficient to compare measured vibration levels with such limits. It is more important to establish the vibration signature of a machine and watch all changes of vibration level during operation, since these are more indicative of deterioration.

The choice of bearing type plays an important role by influencing the mechanical behavior of a turbocompressor. Most machines are designed to accommodate either lemon-type or multisegment bearings with fixed or tilting pads (Fig. 15). The latter are generally used for high rotating speeds, above 10,000 rpm, or for compressors where the bearings are poorly loaded, or machines operating with high-density gases. At high operating pressures a slightly uneven pressure distribution around the rotor can easily cause large radial forces which are much better damped with tilting-pad bearings.

Axial Thrust and Rotor Coupling

The axial thrust exerted on the compressor rotor by the pressure of the gas is almost entirely compensated for by a balance piston at one or both ends of the shaft (Fig. 16). The piston is provided with labyrinth strips rotating with small clearances against a stationary ring or stationary labyrinth.

The remaining thrust is absorbed by a thrust bearing which may be located in one of the bearing housings of the compressor. It is general practice to use bearings with self-equalizing pads which assure an even load distribution (Fig. 17).

When all the machines of a single train are equipped with their own thrust bearing, flexible gear-type couplings are used for connecting the individual shafts. Couplings of this kind are widely accepted since they allow some shaft misalignment without any ill effect. However, the major drawbacks of gear couplings are the associated torque lockup problems and also their susceptibility to damage if the lube-oil passages become blocked by dirt.

Neither of these problems exists with solid couplings. Solid couplings transmit axial thrust. The number of thrust bearings is thus reduced and the mechanical losses minimized. They also allow equalization of the axial thrusts between different rotors arranged for balance of opposing forces.

Unfortunately, very few manufacturers have extensive experience with solid couplings. However, diaphragm-type couplings are now available for compressor drive applications. These couplings transmit only minimal axial thrust and lateral bending forces. They do not require lubrication and are well suited for the majority of mechanical drive applications encountered by the petrochemical industry (Fig. 18).

Material Selection

In many chemical applications, special precautions have to be taken by selecting suitable materials, depending on the nature of the gas to be compressed, the operating conditions, and whether the gas is wet or dry.

For most gases the casing and diaphragms present few material problems since stresses in these parts are relatively low, but rotating parts like impellers or blades are more of a challenge.

Steels with 10 to 18% chromium for these parts ensure good protection in many cases where corrosive gases, such as nitrous gases, have to be handled. They also have the advantage that their mechanical strength differs only slightly from that of steels normally used, thus permitting maximum utilization of the compressor.

For cases where the addition of chromium alone proves insufficient, special steels with high chromium and nickel content, and also containing molyb-

FIG. 16. Typical rotor with the balance piston on the high-pressure side. The outboard side is connected to suction; the inboard side is at discharge pressure. The pressure difference acts toward discharge, opposing the impeller thrust. (Courtesy Sulzer Brothers.)

FIG. 17.   Kingsbury-type thrust bearing with self-equalizing pads. (Courtesy Sulzer Brothers.)

denum, have been developed. These steels are very resistant to H_2 attack, but have inferior mechanical strength, necessitating a reduction in the speed of the compressor. They thus make the compressor more expensive in terms of both size and material. Such materials are used for compressors where intercrystalline corrosion or stress corrosion with gas components such as H_2S may occur.

In machines handling hydrogen-rich gases at high pressures and temperatures, diffusion of atomic hydrogen in steel leads to hydrogen embrittlement: H_2 combines with the carbon present in the steel and forms methane. The methane molecule, being larger than the hydrogen atom, cannot diffuse away and builds up pressures which may destroy the internal adhesion of the steel. The carbon in the steel must therefore be combined with carbide-forming elements such as chromium, vanadium, and molybdenum.

With intercooled compressors, adequate measures must be taken to avoid corrosion in the stages where the dewpoint is reached. The condensate may combine with gas impurities such as SO_2 to form acids which attack the internal parts. Besides installing efficient separators at the cooler outlets to remove the

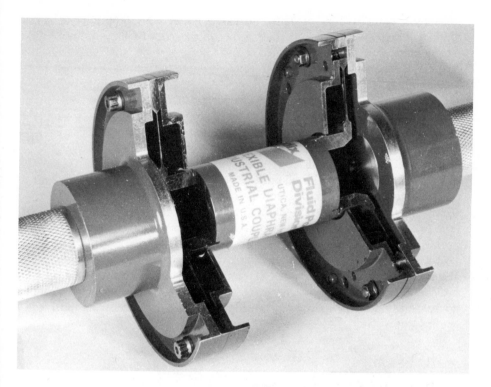

FIG. 18. Flexible diaphragm coupling. (Courtesy Bendix Corp.)

moisture, resistant alloys may also be used. A further measure is to increase the cooling water temperature so that the gas temperature does not drop below the dewpoint. This precaution is particularly important with Cl_2 compressors.

In some cases, added protection against corrosion and erosion can be obtained by using protective coatings.

Injection Devices

When compressing dirty gases or fluids which can cause crystal formation or polymerization, some of these impurities might settle on the inside of the compressor channels and clog the internal passages. This can result in imbalance, reduced flow capacity, increased power consumption, and additional thrust loading due to fouling in balance lines or labyrinths. With nitrous gas compressors in nitric acid plants, nitrate deposits can also present a serious fire or explosion hazard. Injection devices have been developed by several manufacturers for cleaning the insides of compressors, either periodically or continuously, so as to restore the original condition.

Injection nozzles are located in the flow channels, and the washing fluid is injected as close as possible to the deposits. In centrifugal machines, injection is distributed over all stages, and in axials between groups of stages. Nozzles may also be provided to wash the very narrow leakage paths at the rotor seals.

The amount of fluid injected must be governed by the thermodynamic and chemical state of the gas in the compressor. To prevent corrosion of these parts which come in contact with the fluid, high temperatures, high concentrations (due to evaporation of the water), and saturation of the process gas must be avoided.

By using specially designed materials for the internal components, gases can also be compressed fully wet, as in CO_2 compressors for soda plants. Means are provided in the compressor casing for draining the excess fluid, e.g., drain pockets around the periphery of centrifugal diffusors to collect the impurities washed out.

Performance Characteristics

The operating characteristics of a compressor at variable load are usually represented in a diagram of pressure rise against flow volume. Characteristic curves of a centrifugal compressor and an axial machine are shown in Fig. 19. These are valid for given inlet conditions, gas composition, and speed.

A compressor operates at the point where its characteristic intersects the resistance line of the system following it. With small volumes and high discharge pressures, the operating range is limited by the so-called surge line. Beyond this limit the compressor becomes unstable and can no longer overcome the pressure of the system into which it is discharging. A succession of flow reversals, termed surging, then ensues as the compressor alternately discharges gas and the system returns it.

Axial compressors have a steeper characteristic than centrifugal machines. The range of stable flow at constant conditions is smaller, but flow is only slightly affected by variations in discharge pressure.

In the chemical process industry, turbocompressors are required to operate at either constant flow (corresponding to a constant plant output), constant

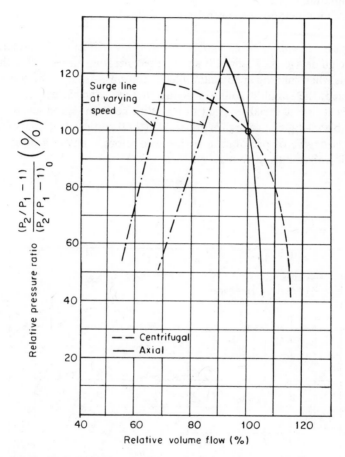

FIG. 19. Performance curves of centrifugal and axial compressors designed for the same operating conditions at the rated point.

discharge pressure (in cases where chemical reactions take place at clearly defined pressures), or constant inlet pressure (to maintain a constant evaporation temperature in a cooling loop). Control can be effected by various methods, but it is important to note that only a single parameter can be varied via operator or instrument decision. The other parameters will find their own corresponding levels.

The main methods of control are as follows.

For *centrifugals*:   speed variation, throttling of inlet flow, and adjustment of inlet guide vanes. Figure 20 shows the operating ranges with the above methods when running at constant inlet conditions. As far as the compressor is concerned, the simplest solution giving the best efficiencies over the widest range is to use a variable-speed drive. With constant-speed electric motors, adjustable inlet guide vanes are preferable to a throttle valve. Inlet guide vanes impart a degree of prerotation to the gas before it enters a given impeller. Throttle valves introduce higher losses.

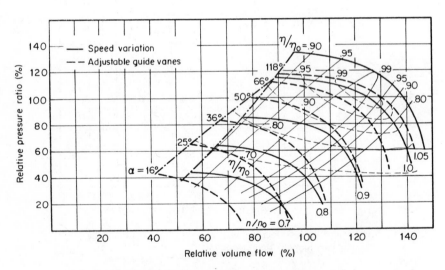

FIG. 20.   Operating ranges of centrifugal compressors with control by variation of speed or adjustment of inlet guide vanes, designed for the same pressure ratio and almost the same maximum suction volume as the compressors in Fig. 21. η is the efficiency, n is the speed, and α is the angular position of the stator blades/guide vanes.

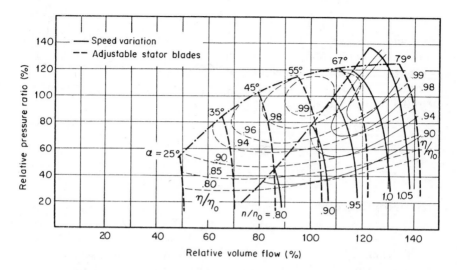

FIG. 21. Operating ranges of axial compressors with control by speed variation or adjustment of stator blades. For the sake of a true comparison, the compressor with fixed blades is designed so that at high suction volumes the two operating ranges almost coincide. η is the efficiency, n is the speed, and α is the angular position of the stator blades/guide vanes.

For *axials*: speed variation and adjustment of stator blades. The respective operating ranges obtained with these two methods are shown in Fig. 21. At constant discharge pressure, the working range of an axial compressor with adjustable stator blades is wider than that of a speed-controlled machine, and the average efficiency is somewhat higher. This is particularly true of compressors having blading with 50% reaction, since higher values are less favorable to angle control. Control by varying the speed may become more attractive if the operating points are all located close to a resistance parabola. However, reducing the speed range for normal operation to a single constant value makes it considerably easier to avoid blade resonance.

By comparing Figs. 20 and 21 it becomes evident that axial compressors with adjustable blades are easily able to match the wide stable operating range of speed-controlled centrifugal machines.

There are two common types of compressor control systems:

1. Pneumatic control with linear servomotors. This system is used for machines of low to medium size and power, or where a control system with relatively slow response is acceptable.
2. Electronic control with high-pressure linear hydraulic servomotors. This form of control is employed for large machines and/or where fast response to the control signal is essential.

Either system can be designed for use in hazardous areas.

Antisurge Control

Unstable running in the form of surging can cause serious damage to a turbocompressor. A reliable control system must therefore prevent the compressor from operating in the dangerous zone.

If the gas demand is at a pressure ratio lower than the surge flow rate corresponding to this ratio, the antisurge controller opens and actuates a blow-off or recycling valve on the discharge side. This results in flow through the compressor greater than the surge flow; the excess is by-passed or vented through this valve (Fig. 22). If the valve is large enough, the compressor can be run at any operating point between the surge limit and zero flow. However, the power expended in compressing the surplus flow is completely lost.

The shape and relative position of the surge line in the compressor characteristic curve is extremely important because the compressor must be protected against surging under all possible inlet conditions and changes in the properties of the gas. The response line at which the controller begins to act is set as close as possible to the surge limit so that the effective operating range is not unnecessarily restricted. Displacement of the surge limit caused by changes in the inlet temperature or the gas constant is allowed for automatically. The control system cannot, however, take into account variation due to a change in isentropic coefficients. In this case the response line must be set in accordance with the worst position of the surge line, i.e., for the lowest values of the coefficient.

With axial compressors the response line of the antisurge controller is also matched to the limits of the rotating stall region which exists near the surge limit at relatively low speeds and flow rates.

In addition to this antisurge protection system, large machines can be provided with an overriding safety device which opens the antisurge valve the moment flow reverses or after one or more surge cycles have been detected.

Antisurge control systems are electronic or pneumatic, as described earlier.

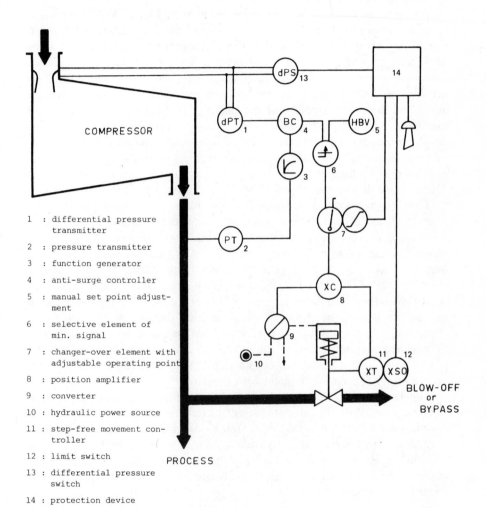

1 : differential pressure transmitter

2 : pressure transmitter

3 : function generator

4 : anti-surge controller

5 : manual set point adjustment

6 : selective element of min. signal

7 : changer-over element with adjustable operating point

8 : position amplifier

9 : converter

10 : hydraulic power source

11 : step-free movement controller

12 : limit switch

13 : differential pressure switch

14 : protection device

FIG. 22. Simplified diagram of an electronic antisurge control system.

Typical Applications of Turbocompressors

Centrifugal and axial compressors are used in *nitric acid plants* for compressing air and nitrous gases to final pressures as high as 14 bar abs. Nitrous gas compressors are made of special noncorrosive materials and are fitted with a water injection system to avoid deposits of ammonium nitrates. A feature of nitric acid units is the use of a tail gas expander. This recovers the energy contained in the residual gases discharged from the absorption column to provide part of the power needed to drive the turbocompressor, while an additional drive unit supplies the balance. For expansion ratios up to 5 and maximum inlet temperatures of 250°C, these machines are of single-stage design with an expander wheel located on the extended compressor shaft. Two- or three-stage expanders are adopted for higher temperatures, up to 550°C. Beyond this, multistage machines are employed up to inlet temperatures of 750°C. Their designs are directly related to industrial gas turbine practice.

Centrifugal compressors are used in *soda plants* to compress the CO_2-rich gases produced in a calcination furnace. Although these gases are being washed and sometimes even electrostatically filtered, they contain a considerable amount of impurities in the form of dust. With dry gases the impurities lead to deposits in the compressor impeller and diffuser passages. The machine therefore has to be opened and cleaned every 2 to 3 months. To avoid this, centrifugal compressors are equipped with water injection at every stage to keep the gas stream completely wet, thus preventing the formation of deposits. Water and silt are continuously evacuated from the return-flow channels. This very efficient system allows uninterrupted operation for 1 to 2 years. Maintenance costs are reduced considerably. Alloy steel and plastic coatings are necessary for parts in contact with the gases.

Ammonia and methanol are produced in synthesis processes operating at high pressures. With the catalysts presently available, a reasonable yield can be obtained with reaction pressures of 180 to 330 bar for ammonia, whereas methanol processes have recently been developed using alternate pressures of 50, 100, or 320 bar. With increasing plant capacities, multistage centrifugal compressors of the barrel design have replaced reciprocating machines. The gases to be compressed are very light. Single-casing units are used in low-pressure methanol installations, from about 150 tons/d capacity upwards, whereas a three-casing arrangement can achieve a final pressure of over 400 bar with a plant output of 1000 tons/d of ammonia or methanol. A recycling stage designed for 5 to 7 times the syn gas capacity is usually located in the high-pressure casing.

Matching high-power, high-speed compressors to steam turbines presents special problems. Where no proven high-speed turbines are available, normal low-speed prime movers and an intermediate gear can be used.

With *catalytic cracking* processes, large quantities of air at pressures of about 3 to 3.5 bar are required to burn-off deposits in the regenerator. Axial machines for the Houdry process were applied as early as 1936. Today, the use of axials for modern cat crackers is becoming increasingly popular with unit powers up to 16 MW.

On the other hand, almost every refinery is equipped with a *reforming* plant, requiring compressors for recirculating hydrogen-rich gases. The compressors used are of the barrel type, even for relatively low operating pressures of 20 to 50 bar, to avoid the possibility of leakage which may occur with horizontally split designs and very light gases. A number of compressors for conditions of this kind are in service worldwide.

Refinery processes also make extensive use of refrigeration compressors for a variety of low-temperature distillation, alkylation, and dewaxing processes.

Normal air compressors are used in *oxidation processes.* They can be of centrifugal or axial design. In acrylonitrile plants, for example, small axial machines handling air flows from about 45,000 m^3/h and discharging at 3 to 3.5 bar abs are used successfully. Uncooled centrifugal compressors as shown in Fig. 23 are widely used for smaller capacities. Similar machines are also in service in maleic or phthalic anhydride plants.

Crack gas and flue gas compressors are used in olefin plants for producing acetylene, ethylene, propylene, or butadiene. They handle a hydrocarbon mixture discharged from the cracking tower. This gas is generally wet, contaminated, and highly corrosive, demanding continuous cleaning of the internal compressor paths by injection, as well as careful selection of materials in order to avoid erosion and corrosion. Cracked gases are usually compressed to about 25 to 35 bar. This is to allow separation of the various constituents of the hydrocarbon mixture. Extensive intercooling is needed because of the danger of polymerization at higher temperatures.

The high vacuum on the suction side in acetylene and butadiene plants results in large suction volumes. As well as supplying installations with large centrifugal machines, some manufacturers have introduced flue-gas compressors of axial design, and these have proved quite reliable under arduous conditions. Their blading is made of special corrosion-resistant steel.

Refrigeration compressors find wide application in olefin plants for separating the cracking products by fractional distillation.

The recovery of ethylene, for example, requires propane (or propylene) and ethylene refrigeration loops in cascade, operating at temperatures as low as $-120°C$.

In other petrochemical plants, such as for the production of synthetic rubbers, large centrifugal ammonia compressors have been installed. The availability record of these units is outstanding in many cases.

Natural gas is liquefied in order to allow shipment in liquefied methane carriers. In this field, one manufacturer has introduced high-power axial

FIG. 23.   Uncooled centrifugal compressor. (Courtesy Sulzer Brothers.)

FIG. 24. Train of oxygen compressors for 2000 ton/day oxygen plant. (Courtesy Sulzer Brothers.)

compressors as refrigeration machines for compressing a mixture of refrigerant hydrocarbons. The final pressure is up to 39 bar. The total power per train is about 78 MW. Control is by speed variation and by adjusting the stator blades. The shaft seals are of the standard floating-ring type with stand-still seals.

Smaller centrifugal compressors are more conventionally used for liquefying natural gas. They recycle nitrogen or hydrocarbon refrigerants, depending on the process.

Gas reinjection is very often used in the underground storage of natural gas or for pressurizing oil fields. This again is a typical application for barrel-type compressors since discharge pressures are usually high.

Process gas compressors or circulators handle the basic gases produced in olefin plants, such as ethylene, propylene, and other hydrocarbons. These are then passed to a variety of chemical processes that yield the final products.

Examples of such compressors are the single-stage recycling machines employed in a number of ethylene oxide processes. The gas must be entirely free from contamination of oil, which is also a very common requirement for processes using a catalyst. The compressors are, therefore, equipped with special hydraulic seals, using water as the sealing agent.

Air and oxygen compressors are used for coal conversion. Coal, high-pressure steam, and oxygen are necessary for the gasification of coal as a first step towards the transformation of coal into liquid fuel. Figure 24 shows one train of oxygen compressors for a 2000 tons/day oxygen plant. Total plant output 24,000 tons/day.

High-Speed, Low-Flow Centrifugal Compressors*

HEINZ P. BLOCH

High-speed, low-flow (HSLF) centrifugal compressors have been successfully applied to compression services traditionally reserved for reciprocating machines.

The centrifugal compressor design eliminates compressor valves and reciprocating forces, and considerably reduces the number of parts with sliding contact or close clearances. It has generally been found that compressor availability will be increased and maintenance costs will be decreased if centrifugal compressors replace reciprocating compressors. Operating costs may be higher for the high-speed, low-flow centrifugal compressors due to generally lower efficiency. However, this disadvantage must be weighed against the generally lower equipment and erection cost of HSLF compressors. These machines are significantly smaller than equivalently rated reciprocating compressors.

Presently available HSLF compressors are designed with only one impeller per compressor casing. This design has been selected because it meets the mechanical limitations of the power transmission and provides a simplified

*Based on sales literature provided by Sundstrand Corp.

49

construction. Maximum power capability is 400 Bhp. Higher power is not possible at this time due to the limitation of the existing speed-increasing gearbox designs.

In addition to the power and head limitations, several other parameters limit the range of performance capabilities of the HSLF compressor. Suction pressure is limited to a maximum of 1000 lb/in.^{2} gauge due to thrust considerations. Discharge pressure is limited to 1400 lb/in.^{2} gauge due to the casing design. Impeller tip speeds are limited to 1300 ft/s due to strength considerations or a Mach number of 1.3 to avoid high losses. These limitations of the HSLF compressor are more clearly presented in terms of their influence when different gas conditions are considered.

If a specific heat ratio (k) of 1.4 is assumed, the maximum pressure ratio which can be developed by the HSLF compressor for gases of different molecular weight and suction temperatures is shown in Fig. 1. The maximum pressure ratio for gases with molecular weights less than approximately 45 is determined by the 28,000-ft adiabatic head capability of the impeller. For higher molecular weight gases, the head is limited by the impeller tip Mach number. That is, the sonic characteristics of these gases require the impeller tip speeds and, therefore, the head capability of the impeller to be reduced.

Performance of the HSLF centrifugal provides a wide range of capacity control for small changes in head rise. This is generally suitable for single and series compressor installations. Parallel compressor installations, however, are not suited to the HSLF partial emission centrifugal performance curve due to the extremes of unbalanced load which can result. Full emission, parallel operation is feasible.

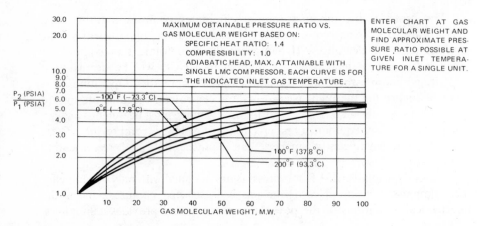

FIG. 1. HSLF compressor performance capability for maximum pressure ratio. (Courtesy Sundstrand Corp.)

The maximum driver size of 400 hp imposes additional limitations on the performance range of the HSLF compressor. Assuming a typical 60% adiabatic compression efficiency and a driver size of 400 hp, the pressure ratio capabilities of the HSLF compressor for different inlet flow capacities and suction pressures are as shown in Fig. 2. As the gas flow increases, the pressure ratio capability is reduced. At constant flow capacity the pressure ratio capability is decreased as the suction pressure is increased. This represents a balance of the pertinent parameters to produce a 400-hp driver requirement at 60% adiabatic efficiency.

The curves of Figs. 1 and 2 represent approximate HSLF compressor applications. If a specific compression requirement is below the respective curves for inlet temperature and pressure ratio, it is likely that an HSLF centrifugal is applicable to the service. Corrections may be necessary for different specific heat ratios and compressor efficiencies.

A cross section of the Sundyne HSLF centrifugal compressor is shown in Fig. 3. The unit is composed primarily of three parts: the driver, which is usually but not necessarily an electric motor; the gearbox; and the compressor or fluid

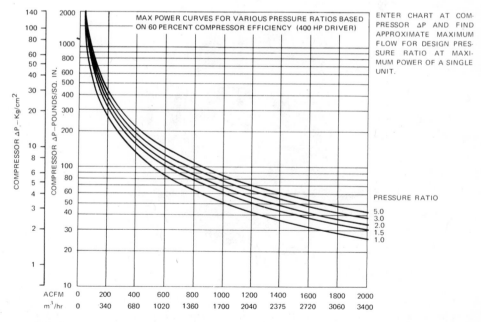

FIG. 2.   HSLF compressor performance capability for power estimation. (Courtesy Sundstrand Corp.)

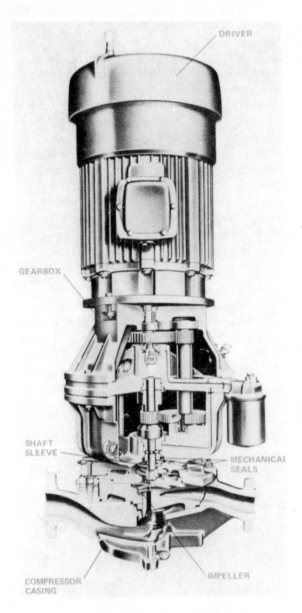

FIG. 3. Cross section of Sundyne HSLF centrifugal compressor. (Courtesy Sundstrand Corp.)

end designed 'for in-line installation. The induction motor drivers available from the manufacturer are operating at nominal speeds of 3000 and 3600 rev/min (50 and 60 cycle operation, respectively). Steam turbine drivers are feasible and have been applied.

The gearbox, which converts the high torque, low-speed input power to the low-torque, high-speed output power, is the key mechanical component which led to the development of the HSLF compressor.

The fluid end, or compressor section, of the HSLF compressor was designed to match the capabilities of the available gearbox. This resulted in impellers with maximum diameters of approximately 12 in.

Standardization and parts interchangeability are notable mechanical characteristics of the HSLF compressors manufactured today. Performance capabilities are presently limited by several of the mechanical components.

Positive Displacement Compressors

RALPH JAMES, Jr.

Reciprocating Compressors

General, Piston-Type

Reciprocating compressors are the oldest and most widely used type. Compression is by the forced reduction of gas volume by the movement of a piston or plunger in a cylinder. Suction and discharge valves are spring loaded and work automatically from pressure differentials generated between the cylinder and piping by the moving piston.

The reciprocating compressor can be obtained in both air-cooled and water-cooled models. Since we are discussing the compressor as related to petrochemical processes, we shall concern ourselves only with the liquid-jacketed type. Reciprocating compressors have a wide range of applications. Speeds may range from 125 to 1000 rev/min. Piston speeds range from 500 to 950 ft/min, the majority being 700 to 850 ft/min. The nominal gas velocity is usually in the range of 4,500 to 8,000 ft/min, and operational discharge pressures may vary from vacuum to 50,000 lb/in.^{2}.

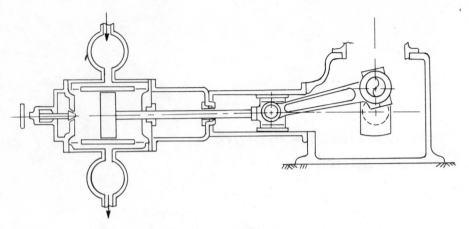

FIG. 1.   Double-acting reciprocating compressor section.

Nomenclature

Crankcase

The crankcase is a U-shaped cast iron or fabricated steel frame. The top is left open for installation of the crankshaft. To prevent the top from opening and closing from the forces of the throws, it is held together with torqued bolts and spacers, or alternately keyed spacers. These are placed directly above the main bearings. The main bearings, spaced between each throw, have removable top covers for removal of the babbitted bearing liner shells. The keyed spacers are, therefore, preferred since their removal is easiest for access to bearing covers. Main bearings, however, in the types of compressors used in petrochemical plants, are so overdesigned for stress that they seldom require removal for rebabbitting.

Crankshaft

The crankshaft is the heart of the machine, and usually the most expensive component. Each throw is forged and counterweights bolted on to balance the reciprocating mass of the crosshead and piston. If the crankcase moves on the foundation, it will cause the throw to open and close through each revolution, and fatigue and break. For this reason the dimension at the open end of the throw must be taken periodically while barring the crankshaft through 360°. This is called taking crankshaft deflections and is recommended as an annual check.

Connecting Rod

The connecting rod is provided with pretorqued bolts fastening cap to body at the crank end. The split is shimmed for removal with wear. The wrist pin is free floating and held in place with caps in the crosshead, which allows the connecting rod to find its own center.

Crosshead

The crosshead runs between two guides with about 1 mil/in. of diameter clearance. It is often weighted so that the mass inertia of all reciprocating parts is sufficient to reverse the stress on the wrist pin, even when one end of the piston is under pressure. If this is not done, the wrist pin will wipe all the oil from the side under stress and will bind.

Lubrication

Lubrication of the frame is accomplished either by a pump driven from the crank end or by a separately mounted pump. It takes oil from the crankcase sump and pumps it through a cooler and filter, usually of 25 μm, and then through piping to the main bearings. The crankshaft has holes drilled from the main bearing surface through to the connecting rod-bearing face. From here, the oil passes up through a hole in the connecting rod to the wrist pin and from there through holes to the crosshead sliding faces. Oil scraper rings in the frame end prevent oil leakage along the piston rod. Because of this tortuous passage of oil, prelubrication is required before start-up. This is accomplished with an auxiliary lube pump. Crankcase oil heaters are specified for outdoor compressors to keep the oil at the required viscosity and to prevent condensation with resultant corrosion. Oil, however, is a poor conductor and local overheating and carbonization has occurred in the use of these heaters. Therefore, when using crankcase heaters while a compressor is not in operation, the auxiliary lube pump should be continuously run.

Cylinder

Materials. Up to 1000 lb/in.^{2}, cylinders are normally cast iron. Above this working pressure the materials are cast steel or forged steel, at the manufacturer's discretion. Nodular iron castings are sometimes specified in preference to cast iron.

API-618 specifies that all cylinders have replaceable liners. These are usually of cast iron because of its lubricating and bearing qualities. Liners should be honed to a finish of 10 to 20 μin.

Cylinder Sizing. Cooling jackets, lubricant and packing, and ring material limit cylinder temperatures. 275°F is set as the ideal maximum with 375°F as the absolute limit of operating temperatures. Using the above limits, the design ratio per cylinder can be set. The design compression and tension load on the piston rod must not be exceeded and, therefore, rod load must be checked on each cylinder application. Rod load is defined as

$$\text{R.L.} = P_2 \times A_{\text{HE}} - P_1 \times A_{\text{CE}}$$

where R.L. = rod load in compression in pounds
  A_{HE} = cylinder area at head end in any one cylinder
  A_{CE} = cylinder area at crank end, usually $A_{\text{CE}} = A_{\text{HE}}$ − area rod
  P_1 = suction pressure, lb/in.^{2}abs
  P_2 = discharge pressure, lb/in.^{2}abs

The limiting rod loads are set by the manufacturer. Some manufacturers require lower values on the rod in tension than in compression. For this, the above limits are checked by reversing A_{HE} and A_{CE}. Note, however, that Exxon Basic Practices impose a limit on the maximum allowable stress acting on the net thread root area of piston rods. This limit is largely based on Exxon's field experience and relates to rod attachment limitations rather than stress failures of the rod itself.

Compressors below 500 lb/in.^{2}gauge, such as air compressors, are usually sized by temperature. Rod load usually becomes the limiting factor on applications above these pressures.

Cooling. In process compression, higher pressures are normal. Therefore, compression ratios are low. Low compression ratios give low temperature rise. Thermosyphon cylinder cooling is, therefore, specified with a discharge temperature limit of about 200°F. Thermosyphon cooling consists of filling the jackets with an appropriate liquid such as water or light oil and letting the heat radiate from the outer cylinder walls. The purpose of filling the jackets is to obtain an even heat distribution throughout the cylinder.

Above 200°F, coolant circulation is applied. Raw water is to be avoided because it leaves deposits in the cooling jackets which are extremely difficult to clean out. A closed system is specified, therefore, which consists of a reservoir, circulating pumps, and heat exchanger.

Overcooling of the cylinder to a temperature below the dew point of the compressed gas must be avoided to prevent cylinder corrosion, lubrication washing, or liquid build-up resulting in a slug. Therefore, the coolant must be by-passed around the exchanger on temperature control. It is also recommended that a thermostat and heater be mounted on the reservoir and the coolant circulated when the compressor is stopped on standby in order to maintain the cylinders at a temperature above the gas dew point.

The coolant is usually 50% glycol and water to prevent freezing. This mixture has a lower specific heat than water. Therefore, the manufacturer must always be asked to size heat exchangers on this basis, even if the initial coolant is intended to be only water.

Valves. Compressor valves are the most critical part of a compressor, and generally have the most maintenance of any part. They are sensitive both to liquids and solids in the gas stream, causing plate and spring breakages. When the valve lifts, it can strike the guard and rebound to the seat several times in one stroke. This is called valve flutter and leads to breakage of valve plates. Light molecular weight gases such as hydrogen are the main cause of this problem which is controlled in part by restricting the lift of the valve plate, thus controlling valve velocity. API valve velocity is specified as

$$V = \frac{D \times 144}{A}$$

where V = average velocity in ft/min

D = cylinder displacement in ft^3/min

A = total inlet valve area per cylinder, calculated by valve lift times valve opening periphery times the number of suction valves per cylinder in $in.^2$

Compressor manufacturers object to the API valve velocity since it gives a valve velocity for double-acting cylinders one-half the value of equivalent single-acting cylinders. Therefore, manufacturers' data on double-acting cylinders often indicate a valve velocity double the API valve velocity, and care must be taken to know the basis upon which valve velocity is given. For heavier molecular weight gases ($M = 20$), API valve velocities of about 3580 ft/min are

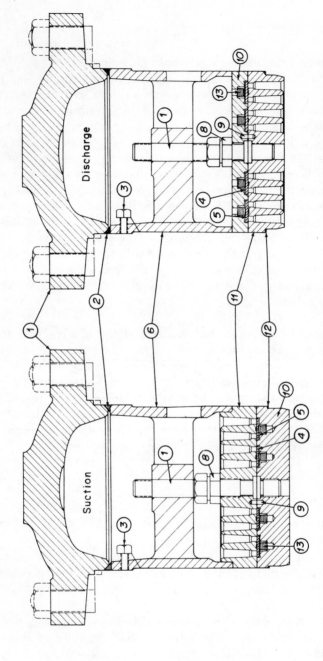

FIG. 2. Single deck plate valves (API). (1) Valve cap. (2) Valve cap gasket. (3) Set screw. (4) Valve plate (inner). (5) Valve plate (outer). (6) Valve cage. (7) Valve stud. (8) Drake locknut. (9) Seat to guard dowel. (10) Valve guard. (11) Valve seat. (12) Valve seat gasket. (13) Valve plate spring.

selected, and for lighter molecular weight gases ($M = 7$), API valve velocities of 7000 ft/min are used.

Manufacturers often use interchangeable suction and discharge valves. This can lead to putting valves in the wrong port, which can result in massive valve breakages or broken rods or cylinders. We specify that valves must not be interchangeable. However, because this feature can be lost or broken off, correct valve placement should always be checked.

Pistons. Pistons are usually cast iron and are often hollow to reduce weight. This space can fill with gas and is an explosive hazard when the piston is removed from the rod. We therefore specify that an easily removable plug must be supplied to vent this space before handling. Larger pistons are aluminum to reduce weight. These pistons have large clearance in the bore to allow for thermal expansion, in the order of 20 mils/in. of diameter. Rider rings are often supplied on CI pistons and are necessary on all aluminum pistons. These, or the whole piston, are rotated 90° about once per year to reduce wear. Bearing loads of pistons and rider rings are based on a unit stress calculated from half the rod plus piston weight in pounds, divided by the diameter of the piston or rider ring, times the width in square inches. Normal limits are 5 lb/in.^{2} for Teflon, 12 lb/in.^{2} for cast iron, 14 lb/in.^{2} for bronze, and 22 lb/in.^{2} for Allen Metal. Teflon compression rings are specified and are useful up to 500 lb/in.^{2} ΔP. Above these loadings, copper-bearing material or babbitt are used for wearing rings and bronze for compression rings. Teflon compression rings are often offered with steel expander rings underneath. These should be avoided because when the compression rings wear, the expander rings can score the cylinder. Designs of pistons and rings are available which will hold the compression ring out against the cylinder without expander rings.

Piston Rod. The piston rod screws into the crosshead, and must be locked, either by a locknut or a pin, to prevent backing off. The rod is adjusted in the crosshead to equalize the end clearance of the piston in the cylinder. This is checked by barring over the machine, crushing a piece of soft lead, and measuring the remaining thickness. This is called the bump clearance.

The rod must be hardened where it passes through the packing. Some rods are chrome plated, but problems have occurred, especially on high-pressure machines with a high heat buildup, causing spider web cracks in the chrome which in turn can flake off and destroy the packing. The best arrangement is to purchase a flame-hardened rod which when worn can be plated with tungsten carbide which should last the life of the machine. High-pressure machines often have the rod extended through the piston and out the cylinder head to balance the pressure load on the piston, i.e., the rod load. This is called a tail rod. Tail rods have been known to break off and fly out of the cylinder like a missile. We specify that all tail rods must be housed in a container strong enough to contain the tail rod should a breakage occur.

Packing. Compressor packing is made up of two rings in pairs, mounted in steel or cast iron cups with the open end of the cups facing away from the pressure. The cups are bolted together, and have vent, oil, and drain holes drilled in them where required. The holes must be correctly aligned each time the packing is opened. If Teflon is specified for packing for pressures above 500 lb/in.^{2}, an additional metallic back-up ring is generally used to prevent the Teflon from extruding out of the cups.

Compressor manufacturers supply a distance piece between the cylinder and the crankcase for access to the packing. This is usually good for a three of four cup set. In process applications, however, especially above 2000 lb/in.^{2}, packing sets run to 8 or 18 cups or more. Extra long distance pieces must be specified in order that sufficient space is available to open the cups and change the packing rings. If any doubt exists, the extra long distance piece should be specified because the extra cost is minor, but it is almost impossible to change the distance piece after the machine is built.

Gaskets. Cylinder end covers, valve covers, and valves are gasketed to the cylinder. Metallic gaskets are specified. Manufacturers often offer fiber gaskets or asbestos-filled metallic gaskets, which have given considerable trouble with leaks, especially on low molecular weight gases.

The gasket seats must often be lapped to obtain a good seal. Soft iron or metallic V ring sets have given the best service. "O" rings confined four ways on valve cover plugs have also given good service, but should never be used where they can be crushed when pulling down the cover. Plug holes must be chamfered to prevent cutting of "O" rings upon entry.

Comparison between Reciprocating and Centrifugal Compressors

In the final analysis, economic factors will govern the selection of a compressor. Both the user and compressor manufacturer should have full knowledge of operating conditions and factors governing selection. In this section, selection factors relating to reciprocating and centrifugal compressors will be covered. As energy costs increase, the higher efficiency of the reciprocating compressor becomes increasingly important, and in some services is now being more often selected.

In some instances, certain factors dictate the use of one or the other machine without question. In other cases all factors must be analyzed carefully in order

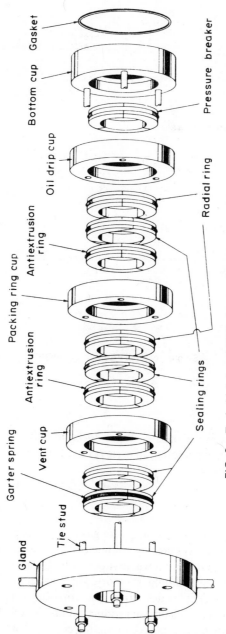

FIG. 3. Typical lubricated TFE piston rod packing, exploded view.

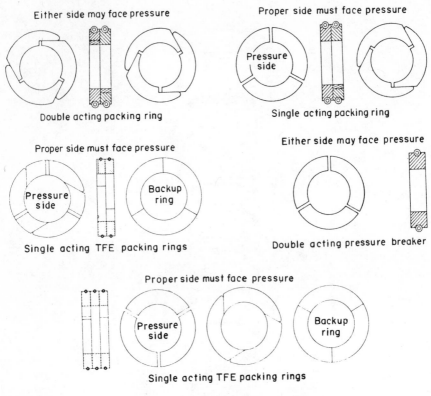

FIG. 4. Typical piston rod packing rings.

to make a choice between the two types. Occasionally, a reciprocating and centrifugal compressor operating in series is indicated.

Gas Properties

Gas Analysis.   A complete and accurate gas analysis is the starting point in compressor selection. Percentage by volume of component gases, entrainment of liquids and solids, and percent water vapor should be recorded. Even minute quantities of contaminants should be reported in the analysis. Trace amounts of sulfur compounds and chlorides can cause corrosion or other mechanical difficulties. Even slight corrosion can produce failure of cyclically stressed parts of either type of compressor. In a reciprocating compressor, solid particles will cause high maintenance costs by accelerating wear of valves, pistons, cylinders,

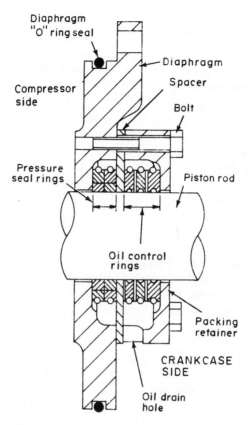

FIG. 5. Typical crosshead diaphragm packing.

piston rods, and packing. Solids passing through a centrifugal compressor may erode impellers and casings severely. If at all possible, any solid particles should be removed from the gas stream before the gas reaches the compressor. In some cases a wash fluid may be used to carry fine solids through a compressor. Water vapor or other vapor in a gas uses up compressor capacity. This volume must be allowed for in sizing the compressor.

Molecular Weight.   Except for a change in efficiency due to larger or smaller valve losses, the reciprocating compressor is not affected by the molecular weight of the gas. Periodic changes in gas composition will have little effect on compression horsepower and pressure. On the other hand, the pressure developed by a centrifugal compressor at a particular speed is directly proportional to the density or molecular weight.* Also, the internal flow passages of a centrifugal compressor are designed to best handle the change in density of a particular gas as it passes through the compressor.

For equal impeller peripheral speeds it will take many stages compressing hydrogen to equal the pressure created by a single-stage compressing a dense gas such as Freon.

If the percent of constituents in a composite gas varies from time to time, so will the molecular weight. A centrifugal compressor, to generate the desired pressure when handling the full flow of such a gas, would have to be designed for the lowest expected molecular weight. Excess pressure resulting from increased molecular weight would have to be reduced by throttling, change in speed, or by some other means. In other words, it is difficult to design a centrifugal unit which will economically and practically handle changing densities.

Polytropic Exponent.   As a gas is being compressed, the polytropic exponent determines the pressure–volume relationship and temperature change. If the polytropic exponent is not known, the average ratio of specific heats may be used for calculating the theoretical adiabatic compression temperatures, volumes, and horsepower.

Temperatures are important in a reciprocating compressor. Temperatures within a conventionally lubricated cylinder should not exceed 350°F.

Theoretically, a gas with a low exponent can be compressed with much higher ratios of compression per stage, yet hold temperatures within desired limits. One drawback to this, however, is that volumetric efficiency decreases with high ratios, and a low exponent may make it uneconomical to compress to high ratios.

The compression exponent also influences the design of centrifugal compressors. Pressure developed by an impeller will be less for a high-exponent gas than for a low-exponent gas of equal density at inlet conditions.

Process Conditions

Flow Rate.   The centrifugal compressor is essentially a large-capacity machine. If 100,000 ft³/min of gas is to be compressed from near atmospheric suction conditions, a centrifugal compressor would probably be used for the lower stages of compression. If the final discharge pressure is high, the last stages of compression could be by reciprocating compressors.

---

*Centrifugal force = [(mass) (tangential velocity)]/(radius). The impeller of a centrifugal compressor will develop the same head for various gases but the pressure is a function of gas properties.

In general, reciprocating compressors have a cost advantage up to 10,000 ft^3/min capacity. Another rough rule of thumb for process applications is that comparison of centrifugal compressors with reciprocating machines is recommended if the volume of a gas stream at discharge pressure is over 600 ft^3/min. Also, custom-built centrifugal compressors are generally a parity investment with reciprocating compressors if the brake horsepower is 2000.

If the flow rate varies widely, the reciprocating compressor can operate with a reasonable sustained efficiency. Flow variations may be handled by suction valve unloaders, clearance pockets, or speed changes.

A suction valve unloader operates by depressing the valve feathers or plates to hold the valve open during both suction and compression strokes. So, on the compression stroke, the gas is not compressed, but flows back into the suction manifold. The unloader can be operated either manually or automatically.

Clearance pockets are volume chambers built into the cylinders or heads or may be attached to the cylinder by a piping connection. Ordinarily a built-in pocket is fitted with a plug valve so that the additional volume may be added or removed at will, either manually or automatically.

Generally, clearance pockets are more flexible in application than are valve unloaders. They can be sized to accomplish any degree of per-stage unloading. Two or more smaller pockets commonly are placed in one cylinder end. Some work is being done on variable-volume clearance pockets. Variable-volume clearance pockets plus suction-valve unloaders permit continuous (infinitely variable) capacity control with minimum power consumption.

With the centrifugal compressor, operation cannot be obtained below the surge point which ordinarily is between 50 and 75% of rated capacity.

Inlet and Discharge Pressure. In a multistage reciprocating compressor a variable suction pressure, while maintaining a constant discharge pressure, may make necessary pockets and/or valve lifters in order to maintain satisfactory operation. Lowering the suction pressure will lower the overall horsepower, lower the differential pressure on all stages except the last stage, increase the differential on the last stage, and, very often, increase the horsepower on the last stage. Raising the suction pressure will raise the horsepower of the complete machine, raise the differential pressure on all stages up to the last stage, and probably lower its horsepower.

In centrifugal compressors an increased suction pressure will raise the discharge pressure and increase the horsepower. If the suction pressure is lowered, the centrifugal machine will not compress to the desired discharge pressure.

Temperature. In both reciprocating and centrifugal machines the compressor recognizes only capacity at the actual inlet conditions. Therefore, the inlet

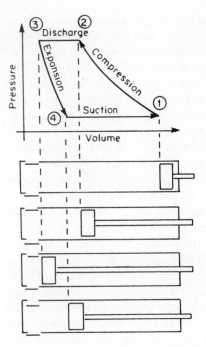

FIG. 6. Ideal $P-V$ diagram for reciprocating compressor. On double-acting compressors the cycle occurs once on each side of the piston for each revolution of the crankshaft. Single-acting machines compress on only one side of the piston. (1) Start of compression stroke, both valves closed. Piston is at bottom dead center. (2) Discharge valve opens and compressed gas is discharged from the cylinder between Points 2 and 3. (3) Discharge valve closes when the piston is at top dead center. Trapped gas at discharge pressure expands to suction pressure as the piston moves back. (4) Suction valve opens. Gas is drawn into the cylinder as the piston continues toward bottom dead center.

temperature must be specified, and compressors are usually rated at suction conditions with conversion to standard conditions for reference and comparison only.

Insofar as mechanical operation is concerned, the centrifugal compressor is less affected by high or low temperature extremes than is the reciprocating compressor. Centrifugal compressors have been used to circulate gas at 800°F. With conventional lubricants such temperatures are impractical in a reciprocating compressor. Very low temperatures also cause lubrication problems. However, reciprocating compressors have been operated at suction temperatures below $-100°$F.

A nonlubricated reciprocating compressor which has been developed in Europe may solve both these lubrication problems. A grooved piston which does not contact the cylinder walls is used. Close clearances are employed to control leakage between the cylinder and piston. Good results have been obtained with this design.

A centrifugal compressor designed for a given capacity at a specified inlet temperature will fail to deliver the required discharge pressure if the suction temperature is increased significantly. Or, if the suction temperature is lowered, the discharge pressure will be increased. Consequently, the centrifugal compressor must be designed to deliver the desired capacity and pressure under the maximum inlet temperature conditions.

Heat Balance.   Process heat balance sometimes has a bearing on the selection of a driver. In turn, the driver selection may determine the compressor type. For example, if a backpressure steam turbine driver is selected because of low-pressure steam requirements, a centrifugal compressor would be a logical choice. Steam turbines are ideally suited for centrifugal compressors for several reasons. First the revolutions per minute match between centrifugal compressors and turbines permits a direct drive. In the size range of centrifugal compressors, turbines are one of the least expensive drivers. They are mechanically efficient and lend themselves well to speed control by which compressor pressure and gas flow may be indirectly controlled, and they rate high on the list for reliability and maintenance cost. They may be used in continuous service of up to 3 years. They are limited, of course, to use in plants where steam is available at reasonable cost.

If economics or other reasons dictate the use of electric motor or gas engine drivers, reciprocating compressors are usually the logical choice. One exception is that gas engines drive large single-stage centrifugals through speed-increasing gears in some gas pipeline stations. This particular combination is less expensive than gas engine reciprocators of the same size.

Under certain conditions, such as low fuel cost plus deficiency in moderate pressure steam, a combustion gas turbine driver may be selected. Credit for the steam produced by a waste-heat boiler will offset the high first cost of the gas turbine.

Series and Parallel Operation

Within certain limits, compressors can be operated in series or parallel. In positive displacement machines operated in series, the discharge flow rate of the first must equal the inlet flow rate of the second. If these are not carefully matched, a series system may produce an appreciable vacuum or excessive pressure between the two compressors.

With turbocompressors in series, a flow mismatch will cause surging of one or the other, and flow instability.

Positive displacement compressors can be operated in parallel if the discharge pressures of the two machines are about the same; flow through one compressor has little or no effect upon the other. However, turbocompressors operated in parallel do present a matching problem; one compressor will frequently pick up the entire flow while the second idles or becomes unstable. Consequently, an automatic flow-control system is recommended.

Turbo and positive displacement compressors can be combined; a turbocompressor can be used in series with a positive displacement machine, or vice versa. A turbocompressor could also be used in parallel with a positive displacement compressor as a flow booster.

Integral Compressor-Engine

Figures 7, 8, and 9 present a typical gas engine-compressor as manufactured by Cooper Bessemer.

The unit consists of a combination V-type gas engine and horizontal compressor built into one compact unit. The engine is of the two cycle type and is built in units of 6, 8, and 10 cylinders with the number of compressor cylinders varying according to requirements and arranged to give any combination of volume and pressure within the rating of the engine.

The power cylinders are arranged in two banks at 60° with respect to each other and 30° on each side of the vertical centerline. In the case of the atmospheric engine, the power piston controls the opening and closing of both the air inlet and the exhaust ports. In the case of the supercharged engine, the power piston controls the opening and closing of the air inlet ports and the opening of the exhaust ports. The closing of the exhaust ports, however, is controlled by a rotary butterfly valve in the exhaust passage of each cylinder. This valve is timed to close the exhaust ports while the air inlet ports are still open. This timing has the combined effect of not only trapping a greater volume of scavenging air, but at a somewhat higher initial pressure, giving a supercharging effect which enables the engine to burn more fuel and carry a great load.

Scavenging air is supplied by horizontal scavenging air cylinders which also furnish the support and drive for the compressor cylinders. The scavenging pistons are attached to conventional crossheads which are driven by master connecting rods. The scavenging cylinders discharge into a common receiver cast in the base from which the scavenging air is admitted to the power cylinders. On the downward or explosion stroke the power piston first uncovers

FIG. 7. Type GMV-10-TF gas engine, operating side.

FIG. 8. Type GMV-10-TF gas engine, compressor side.

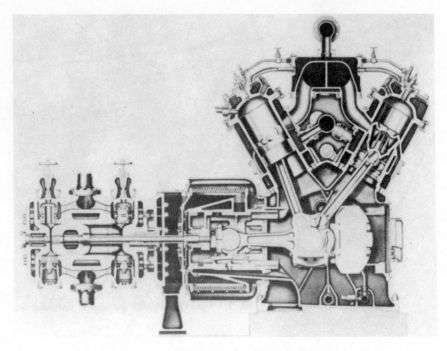

FIG. 9. Type GMV-TF gas engine, cross section.

the exhaust ports, allowing the burned gases to escape to the atmosphere through the exhaust system. Further movement of the piston then uncovers the air intake ports, and scavenging air under pressure in the receiver rushes through the transfer passages into the cylinder, sweeping the remaining exhaust gases out and filling the cylinder with a charge of fresh air. In the case of the atmospheric engine, the power piston on its return stroke first closes the air intake ports and then the exhaust ports, trapping the fresh air in the cylinder to begin the compression stroke. In the case of the supercharged engine, the butterfly exhaust valves close before the piston closes the air inlet ports, thus trapping more fresh air and at a somewhat higher initial pressure before starting the compression stroke.

At about the time the ports are closed, but before the compressor has started, the mechanically operated injector valve at the top of the cylinder opens and a charge of gas under pressure is admitted in an amount regulated by the governor in accordance with the load requirements. The streamlined flow of the scavenging air into the cylinder is so directed as to sweep out the burned gases and mix thoroughly with the fuel gas during the compression stroke to form a very intimate mixture of proper proportions. Shortly before top dead center is reached, ignition takes place, combustion occurs, and the cycle is then repeated.

A layshaft mounted between the two banks of power cylinders is chain driven from the flywheel end of the crankshaft and serves to drive most of the auxiliaries at the opposite end. The air starter, speed regulating governor, lubricators, and magnetos or interrupters are all gear or chain driven off the end of the layshaft. In the case of the supercharged engine, the butterfly exhaust valve shafts are gear driven off the drive end of the layshaft in timed relationship with the crankshaft.

The principal operating parts of the engine including main connection rod, piston pin, crosshead, and layshaft bearings as well as all drive chains and gears are lubricated off the pressure lubricating oil system. All power pistons are jacketed and are oil cooled by lubricating oil from this system. A built-in lubricating oil pump mounted at the operating end and direct driven from the crankshaft furnishes an ample supply of oil at the desired pressure. The oil is cooled and filtered before going to the distribution system. Separate mechanical force feed lubricators supply lubrication to the power cylinders and to the compressors. In the case of the supercharged engine, an independent oil pump and distributor supplies oil to the butterfly exhaust valve shaft bearings.

A low tension ignition system which employs either two magnetos or a single radial-type interrupter with storage battery is used.

Starting is accomplished by means of a built-in air starter using compressed air at 250 lb/in.² maximum pressure.

Other Compressor Cylinder Arrangements

Exact compressor cylinder arrangement varies considerably with individual manufacturers, and only some of the more significant types will be covered here.

Horizontal Balanced Opposed

See Fig. 10 for a typical machine.

The crankshaft extension from one end of the frame provides great flexibility in the use of drivers. A synchronous or induction electric motor, steam turbine (with gear), steam engine, or internal combustion engine may be readily employed.

FIG. 10.   Horizontal balanced opposed multiple cylinder compressor.

Vertical Labyrinth-Piston

A unique design by Sulzer utilizes a contactless seal between piston and cylinder wall and between the rod and gland. This labyrinth seal requires no lubrication. Compression systems wherein small leakages of gas are harmless (air, oxygen, nitrogen, carbon dioxide) are often handled with the open type contractions (see Fig. 11). The totally enclosed model is usually specified for noxious, toxic, or valuable gases.

Hypercompressors*

General

Intimately bound to the chemical industry and based on a long evolution, the main stages of which were the liquefaction of air and the synthesis of ammonia, the technique of using very high pressures was eventually perfected from developments in the manufacture of low-density polyethylene. This is now the only industry requiring large reciprocating compressors for very high pressures, as the pressures necessary for other main chemical processes have been steadily reduced since 1945. For this reason, this description will be restricted to ethylene compressors.

Considering that the classical designs of reciprocating high-pressure compressors cover an uninterrupted range up to about 1000 atm, "very high pressures" will imply those above 1000 atm.

A characteristic feature of the high-pressure ethylene polymerization process is that a very large difference in pressure is necessary between the inlet gas entering the reactor and the outlet of the recycle gas. The recirculators, generally called secondary compressors (Fig. 12), work between two limits, i.e., 100 to 300 atm on the suction side and 1500 to 3500 atm on the delivery side, for most of the existing processes. As the coefficient of reaction lies between 16 and 30%, the secondary compressors have to handle 3 to 6 times the fresh gas quantity, thus being by far the most powerful machines in the production stream. Their unit capacities which, when the industrial expansion first began, were of 4 to 5 tons/h, now lie between 15 and 50 tons/h and their power requirement per unit has been increased from 600 hp up to some 10,000 hp.

As the whole operating range of these secondary compressors takes place well above the critical point of ethylene, the thermodynamic behavior of the fluid lies somewhere between that of a gas and that of a liquid. This peculiar

---

*The following description of very high pressure compressors is adopted by permission from an article by C. Matile entitled *Industrial Reciprocating Compressors for Very High Pressure.*

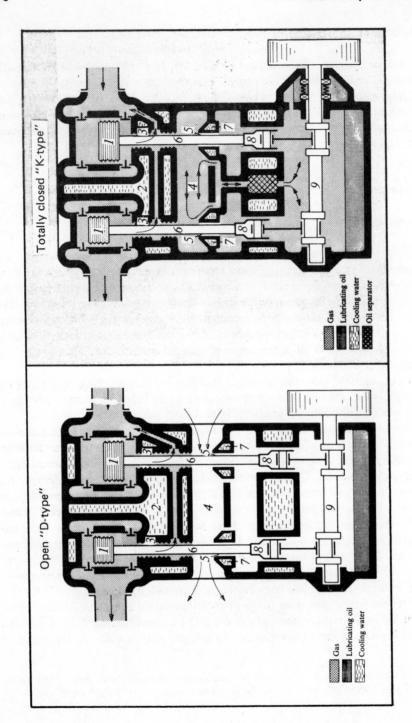

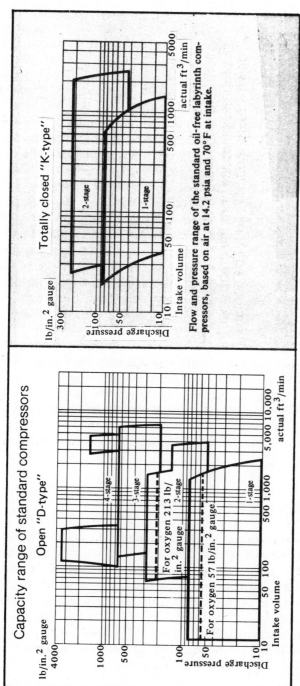

FIG. 11. Labyrinth-piston compressor design. (1) Labyrinth piston: Double-acting, vertical, lightweight design. Exactly centered with very little clearance. Labyrinth surface "grooves" form throttles for blow-by gas. (2) Cylinder: Water-cooled or heated, also provided with grooves on the inside. (3) Gland: A system of graphite rings forming a labyrinth seal. Leaking here is usually led back to the suction side of the compressor. (4) Distance piece: Provides distinct separation between cylinder and crank gear, ensuring that the part of the piston rod covered with a molecular oil film does not enter the gland. (5) Oil wiper: Prevents oil from creeping up the piston rod into the glands and cylinders. (6) Piston rod: Guided very accurately by guide bearing and crosshead. (7) Guide bearing: Oil-lubricated and water-cooled. (8) Crosshead: Oil-lubricated with water-cooled crosshead guides. (9) Crankshaft. Note: The flow and pressure range of the standard oil-free labyrinth compressor is based on air at 14.2 lb/in.²abs and 70°F at the intake.

condition has two main effects. The first is a very small reduction of the specific volume with increasing pressure; for instance, at a temperature of 25°C the specific volume is 3 dm^3/kg at 700 atm and 1.5 dm^3/kg at 4500 atm. The second effect is a very moderate rise of adiabatic temperature with increasing pressure; for instance, with suction conditions of 200 atm and 20°C the delivery temperature will reach only 100°C at 2000 atm.

These particular thermodynamic conditions greatly influence the design of high-pressure ethylene compressors. Compared with the classical reciprocating compressor, the compression ratio is of little practical significance, the important factor being the final compression temperature which should not exceed 80 to 120°C, depending on process, gas purity, catalyst, etc., in order to avoid premature polymerization. The influence of the cylinder dead clearance on the volumetric efficiency is slight because of the small reduction of specific volume, and very high compression ratios are therefore possible with quite admissible efficiency. In addition, the stability of intermediate pressures depends chiefly on the accuracy of temperatures; for instance, in the case of a two-stage compression from a suction pressure of 200 atm to a delivery pressure of 2500 atm, a drop in the first-stage suction temperature from 40 to 20°C will cause the intermediate pressure to rise from 1000 to almost 1600 atm.

For these reasons, a secondary compressor is required which has only one or two stages despite the very large pressure differences involved. However, this again compels the designer to face extremely high mechanical strains due to the high amplitude of pressure fluctuation in the cylinders. Finally, an additional and sometimes disturbing feature of ethylene must be mentioned. If the gas reaches a very high pressure and a high temperature simultaneously (which can easily occur in a blocked delivery port owing to the very low compressibility), it will decompose into carbon black and hydrogen in an exothermic reaction of explosive character.

Cylinders and Piston Seals

Sealing of the high-pressure compression chamber is a major problem, and this could be solved by avoiding friction between moving and stationary parts. This has been realized for laboratory equipment and small-scale pilot plants by the use of either metallic diaphragms or mercury lutes in U-tubes, and such arrangements are still in use for research purposes. In addition, they have the advantage of avoiding any contamination of the compressed gas by any lubricant. Unfortunately, chiefly for economic reasons, they proved to be impracticable for industrial compressors, at least in the present state of

FIG. 12. Large secondary compressor.

techniques. Thus, as labyrinth seals are out of the question for very high pressures, friction seals have to be accepted; in fact, two solutions are currently used—moving and stationary seals.

Metallic piston rings are the only sort of moving seals used in the large reciprocating type of compressor. They are generally made in three pieces; two sealing rings, each covering the slots of the other, and an expander ring behind both of them which also seals the gaps in the radial direction. The materials used are special grade cast iron, bronze, or a combination of both, with cast iron or steel for the expander. The piston, of built-up design, comprises a series of supporting and intermediate rings with a guide ring on top of them and a through-going bolt (two different designs are shown in Fig. 13). All parts of the piston are made of high tensile steel, and particular care must be given to the

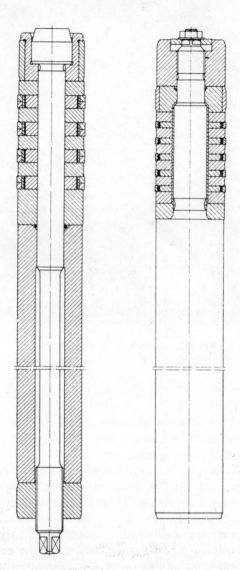

FIG. 13. High-pressure pistons with piston rings.

design and to the stress calculation of the central bolt which is subjected to severe strain fluctuations.

The use of piston rings allows for a simple cylinder design, the main part of which is a liner that has been thermally shrunk to withstand the high variations of the internal pressure (see Figs. 14 and 15). The inner sleeve, which was previously made of nitrided steel, is now generally of a massive sintered material like tungsten carbide. The use of this expensive material is justified by two beneficial qualities; it possesses an extremely hard surface and has a high modulus of elasticity. The first considerably improves the conditions of friction and greatly reduces the danger of seizure. Owing to the high modulus of elasticity, the amplitude of the "breathing" movement under the internal pressure fluctuation is much smaller than with steel, and thus the stress variations in the expanded outer sleeves are appreciably reduced. However, as these sintered materials have a very poor tensile strength, care must be taken to ensure that the inner sleeve is always under compression, even if the temperature increases. This is the main purpose of the external cooling of the liner and not, as is usual, to dissipate the heat of compression.

Packed plungers are the other answer to piston sealing. Although some manufacturers still use packings of hard plastic materials (nylon or similar), the most widely used packings are the metallic self-adjusting type. They are usually assembled in pairs, the actual sealing ring tangentially split into three or six pieces being covered by a three-piece radially cut section. Both are usually made of bronze, kept closed by surrounding garter springs, and held in place by locating and supporting steel plates. These plates must also be thermally shrunk to resist the high variations in internal pressure. Unfortunately, the use of sintered hard materials is restricted by the fact that the supporting plates are subjected, in the axial direction, to heavy bending and shearing forces which these materials generally cannot stand. To improve the friction conditions of the packing rings, the high tensile steel supporting disks are frequently surface hardened or plated with carbide. The plungers are made of nitrided steel for use in moderate pressures, and for high pressures are of steel, plated with hard materials. For very high pressures the use of solid bars of hard metal is the best wear-resistant solution for both plungers and packings. The disadvantage of the packed plunger design lies in the much larger joint diameters of the static forces than the piston ring design. Large cylinders, such as the one shown in Fig. 16, need a pretensioning of the cylinder bolts to about 10 times the maximum plunger load. This ratio is higher for smaller cylinders.

For piston rings and packed plungers the optimum number of sealing elements appears to be four or five. In both solutions it is essential that the piston be accurately centered if the seals are to be effective; this is the reason for the guiding ring within the cylinder and for the additional guide at the connection between the piston and driving rod. At the base of the cylinders an

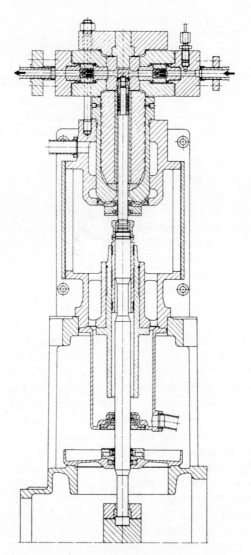

FIG. 14. High-pressure cylinder for moderate end pressures.

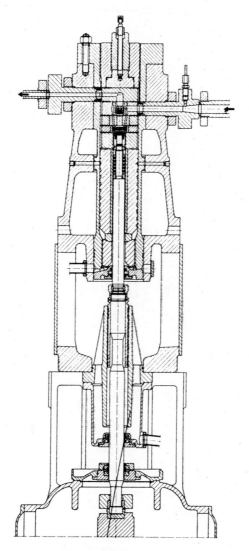

FIG. 15. High-pressure cylinder for medium end pressures.

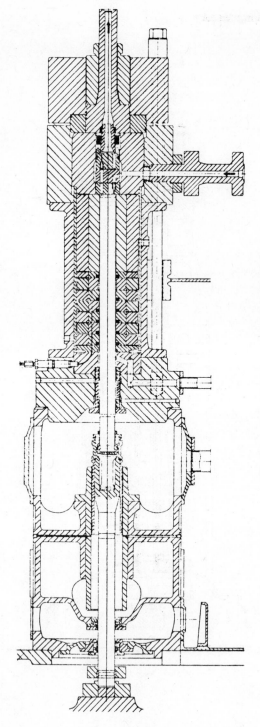

FIG. 16.   Gas cylinder for very high end pressures.

additional low-pressure gland allows gas leaks to be collected and the plunger to be flushed and cooled. Other separate glands positioned on the rod connecting the piston to the drive (see Figs. 14 and 15) prevent the cylinder lubricant from mixing with the crankcase oil, and as the intermediate space is open to the atmosphere, it is impossible for gas to enter the working parts.

From the point of view of design and maintenance, piston rings would appear. to be the most adequate solution, and they are currently used for pressures up to 2000 atm, or in some circumstances up to 3000 atm. The choice between them and the packed plungers depends largely on the process and type of lubricant used. One difficulty is that normal mineral oils are dissolved by ethylene under high pressure to such an extent that they no longer have any lubricating power. The glycerine used in earlier machines has been widely replaced by paraffin oil, either pure or with wax additives, which is much less diluted by the gas than other mineral oils. However, it is a rather poor lubricant and is inferior to the various types of raw synthetic lubricants which are generally based on hydrocarbons. The basic difference between piston rings and plunger packing is that the latter may be lubricated by direct injection while the former are lubricated indirectly. This may be an advantage since the low polymers carried by the return gas back from the reactor are reasonably good lubricants. However, too large an amount of low polymers causes the rings to stick in their grooves, and some kinds of catalyst carriers also brought back by the gas are excellent solvents for lubricants. Thus the most convenient solution has to be selected for each specific case. In general, for higher delivery pressures (above 2000 to 2500 atm), better results are obtained with the use of packed plungers.

Cylinder Heads and Valves

It is relatively simple to construct a vessel that will resist 2000 atm, but the problem becomes more intricate when the vessel must withstand, for years, a pressure which fluctuates between 300 and 2000 atm at a frequency of 3 to 4 Hz. The leading idea of the designer must be to divide a complicated problem into a series of simple ones, each of which is then accessible to accurate methods of investigation. If this is done properly, it is possible to divide a large piece at the very places where inadmissible changes of stresses would occur, and to keep the combined strains in each item within tolerable limits. The examples of cylinders shown in Figs. 14, 15, and 16 show the results of this method. The striking feature is the very simple shape of all pieces subjected to high pressure.

A first obvious result is that suction and delivery valves have to be located in a separate cylinder head. Figure 14 shows one type of cylinder head that can be used for moderate pressure fluctuations (up to amplitudes of about 1200 atm) and moderate cylinder dimensions. The intersection of the gas passages with the main bore is located in a small forged core, shrunk in a heavy outer flange, and pressed by the upper cover in the axial direction. This piece has a quite symmetrical shape with carefully rounded internal edges. By dismantling only the upper cover, it is possible with this design to pull out the complete piston with its rings through the central hole without disconnecting the gas pipes and without removing the valves. A typical valve for this kind of cylinder head is shown in Fig. 17(a). The same valve is used on the suction and delivery side, the two endpieces being differently shaped to avoid incorrect assembly. The valve is held against the central head piece by the connection flange of the gas piping, as a kind of composite lens.

For higher amplitudes of pressure and larger cylinders, cross bores and duct derivations must be taken away from the area of large pressure fluctuations. This is effected by the use of central valves, combining suction and delivery valves into one concentric set. For cylinders of moderate size this can be done as shown in Figs. 15 and 17(b); the different valve elements are located in a succession of simply shaped disks with the same diameter as the cylinder liner and piled up on top of it. The lower two disks, which have been produced by the shrinking technique, receive the pressure fluctuation in their central hole, while the upper two, which contain the radial bores for the gas connections, are subjected only to static pressure.

For still larger cylinders the combined valve is assembled as a separate unit in order to keep compact dimensions and weights, and is inserted into the central hole of the cylinder head core as shown in Fig. 16. The two valves illustrated in Figs. 17(c) and 17(d) are designed on the same basic principle. The last one, used in very large cylinders, is fitted with multiple suction and delivery poppets in order to reduce the moving masses. The entire valve body is subjected to suction pressure on the outside and only to the pressure fluctuations in the longitudinal hole. The suction pipe is connected to the radial bore of the cylinder head core as shown in Fig. 5.

Separation of suction and delivery pressures is assured by the circumferential self-sealing ring of hard plastic material as shown in Figs. 17(c) and 17(d). The entire valve is pressed on the end of the cylinder liner by the difference of pressures; the set of plate springs visible in the figure have only to maintain the valve against pressure drop during periods of operation on bypass. The gas delivery pipe is connected radially to the core piece (like the suction one) for moderate delivery pressures and axially for higher pressures (as shown in Fig. 16).

All components subjected to high stresses, particularly the internal cylinder elements under high tridimensional fatigue strains, are generally investigated at the design stage by three different methods. The first is a conventional calculation of combined stresses based on the classical hypotheses, using

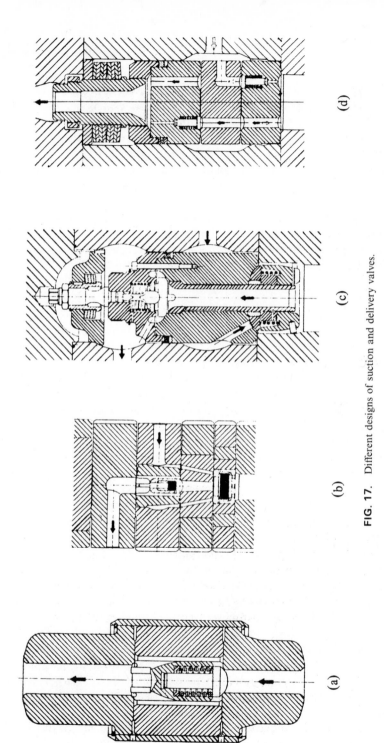

FIG. 17. Different designs of suction and delivery valves.

(a)

(b)

(c)

(d)

87

computer programs as far as convenient. The second approach is that of the frozen stress technique of photoelasticity applied on resin models cast either on full scale or on slightly reduced scale; it supplies accurate information about the course of the two main stresses in every plane section within the material. The third method is a direct measurement of the superficial stresses by means of strain gauges on the actual component subjected to the full prestressing and internal pressure. A variation of this last method consists in stress-measuring on an enlarged model made of a low-modulus material like aluminum; it provides better information through strain gauges on small rounded edges, and allows progressive modification of such places in an attempt to reach an optimum. Comparison of results of these different methods gives a very useful reciprocal check on their exactness and accuracy.

Drive

Different types of driving mechanism are diagrammatically illustrated in Fig. 18. The first two (Figs. 18*a* and 18*b*) have been extensively used during the initial period of development and are still applied to smaller units. They are characterized by the fact that the high-pressure cylinders have been fitted to frames of classical design without substantial modification of the existing equipment. Some manufacturers did not take into account the purely unilateral loading of the crosshead pins—and they generally got into trouble. Others tried to balance the forces by getting additional pistons set under constant or variable gas pressure in the reverse direction; this may work, but it is a rather unsatisfactory solution as it is expensive and introduces supplementary wearing elements. The best means of application is to use special high-pressure lubrication pumps, fastened to the crossheads and driven by the rocking movement of the connecting rods, which inject the oil directly into the crosshead bearings, thus lifting the pins against the load. This arrangement is well known from the design of large diesel engines, but as the requirements called for high delivery pressures and larger capacities, the solutions shown in Figs. 18(*a*) and 18(*b*) appeared to be increasingly unsatisfactory. Since it is impossible to use double-acting pistons on very high pressures, these designs load the driving mechanism with the full gas pressure (instead of the difference between delivery and suction pressures) and work only on each second stroke. While this was still admissible for small units, it proved to be uneconomic for larger ones, and there was obviously a need for more specialized constructions.

The widespread design represented in Fig. 18(*c*) is still based on a conventional application of the horizontally opposed reciprocating compressor, but it avoids the above difficulty by having, on each side of the frame, an external yoke which is rigidly connected to the main crosshead by means of

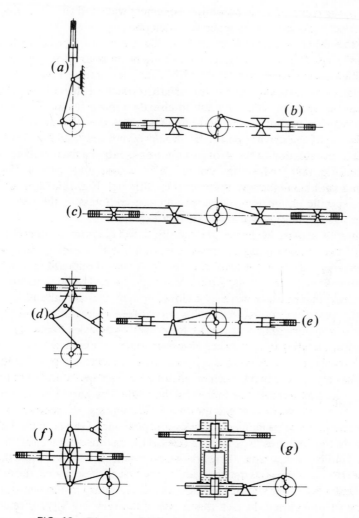

FIG. 18. Diagrams of different types of driving mechanisms.

solid connecting bars. A pair of opposed pistons (or plungers) is then coupled to each yoke which is shaped as an outboard crosshead. This is not a bad solution because, due to the long flexible connecting bars, the movement of the yoke is not disturbed by any transverse force, and it allows a full loading of the drive. However, the compressor is becoming extremely wide and the accessibility to each second high-pressure cylinder is rather poor.

All the other systems illustrated in the figure are specially designed solutions; Figs. 18(d) and 18(f) use a rocking beam bound to a fulcrum by a level, which gives a linear translation of the rotary movement. Figure 18(e) uses a moving frame surrounding the crankshaft to connect the crosshead to the piston on the opposite side—a solution already applied for more than half a century to high-pressure pumps; and Fig. 18(g) is based on the idea of hydraulic transmission of the driving power. It should be noticed that Fig. 18(d) may perform a modification of the stroke of the crankshaft in a fixed predetermined ratio, that Fig. 18(f) reduces the stroke in a fixed ratio, and that Fig. 18(g) can perform a variable reduction of the stroke. Although Fig. 18(e) appears to be the best specific design for a large production compressor, the very special solution of Fig. 18(g) is worthy of further explanation.

Figure 19 shows the basic, greatly simplified diagram of operation. By means of two reciprocating columns of fluid, a double-acting primary piston operates a secondary piston located above it. A pair of opposed high-pressure gas pistons are coupled to the latter. Although the hydraulic transmission of power could theoretically work as a closed system, it is actually necessary to renew the fluid continuously through a forced feed recirculation, both for the purpose of cooling and to compensate for possible seal leaks. Figure 19 shows a low-pressure feeding system; it has also been made as a high-pressure feed. As this transmission may be built as a hydraulic intensifier, it is possible to use a comparatively light primary mechanism at rather high speed, and to reduce the linear speed and increase the forces on the secondary part. Furthermore, by opening a bypass valve between the two fluid columns, the secondary stroke may be reduced. In this manner a stepless output control can be achieved down to zero. As the fluid pressures on both sides of the pistons vary according to two opposed indicator diagrams, there are two points on each stroke where they will balance. If such a bypass is opened wide on the first of these points, the fluid will theoretically flow over without losses, and the secondary piston will stand still until the valve is closed. In fact, this is one of the very few ways of realizing a power-saving capacity control of reciprocating compressors for very high pressures. This output control can be governed automatically, and by being applied separately to each compression stage makes it possible to control exactly the intermediate pressure.

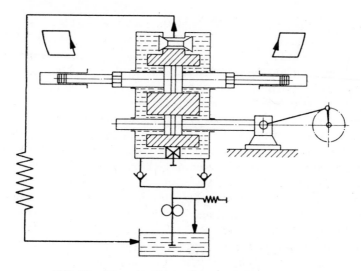

FIG. 19. Diagram of hydraulic transmission of power.

Miscellaneous

The number of problems posed by industrial reciprocating compressors for very high pressures is almost unlimited. Nearly every question of installation or maintenance needs a special study and an original answer, and all elements and accessories require special design and calculation. Only a few of them will be mentioned here.

It is common practice, when designing large reciprocating compressors, to take into account three different kinds of strains for selecting the most favorable crank angle arrangement. First, there are the resulting forces and moments of inertia acting on the foundations; second, the resulting torque diagrams under different conditions of operation (important for the cylic variations of current consumption of the driving motor); and last, the forces due to pressure pulsations in the gas piping. For most compressors working at lower pressures, this last consideration may be deleted or answered in a summary way at the initial stage of design because it may be solved by the use of surge drums. In the case of very high pressures, the gas pulsations, which are capable of destroying the piping system, have to be given the first priority in the basic investigations, even if this sometimes leads to acceptance of higher inertia forces.

The most practical way of studying the gas pulsations is to use an analog computer, which is in fact an electroacoustical analogical system where every part is individually adjustable or replaceable. The first purpose of the analysis is to avoid any resonance between the active systems (the compression cylinders) and the passive systems (the whole piping network); the second purpose is to reduce the amplitudes of the remaining pressure pulsations as far as possible. Theoretically, the means available are: change of diameter and of length of gas piping, removal of pipe connections or adjunction of additional piping, and use of pulsation snubbers and of orifices at well-selected places. In reality, the possibilities are restricted because of the high speed of sound in the gas (1000 to 2000 m/s), because of the very low compressibility of the gas, and because of the very high price of vessels and piping. However, in many cases it has proved easily possible to reduce a dangerous pulsation (for instance, from 25% down to 5%) by cheap and simple means.

Designers dealing with compressors for very high pressures need to keep in mind at least three basic ideas: (1) safety, (2) large forces (how to apply them), and (3) accessibility. The last two are, of course, chiefly economic, but they are often combined with the aim of safety. For instance, in the design of large compression cylinders, as shown in Fig. 5, the long through-going bolts connecting the base with the cylinder head are an important safety factor; if by chance the gas decomposed in the cylinder, these long bolts, acting as springs, would be elastically lengthened by an appreciable amount without significantly increasing the stresses, and would allow the gas to escape between the liner and the head. They must all be equally pretensioned with a very high force. If done by hand, this would be an extremely tiring and time-consuming exercise, and for this reason a hydraulic piston has been incorporated within the cylinder head which allows, when set under oil pressure, a very quick, easy, and regular tightening and loosening of the bolts. After removal of the outer flange, the whole inside of the cylinder can be removed, with the help of a lifting device, as a closed cartridge, as shown in Fig. 9. The same figure also shows how major parts of the driving mechanism may be dismantled without removal of the cylinders.

Conclusion

Although closely related to other reciprocating compressors, the industrial compressors for very high pressures require the construction of a separate group of machines, different in many ways, and call for much greater research, development, and calculation than the others. Being compelled to employ all materials very near to their limits of resistance, the designers are bound to keep in close contact with the latest developments of science and techniques in many fields.

Diaphragm Compressors

General

The diaphragm compressor is a "leak-proof" compressor which uses a piston in hydraulic oil to actuate a diaphragm to displace gas. This positive diaphragm displacement takes in the gas through a suction check valve and discharges a like amount of gas through a discharge check valve. The diaphragm compressor has found extensive application as a booster compressor for flows from 0.3 to 30 actual ft³/min and pressures from 1000 to 50,000 lb/in.² at pressure ratios up to 20/1 in one cylinder. In this application region of small flows and small horsepowers, the conventional reciprocating compressor is not considered competitive as special construction is required. The diaphragm compressor has definite limitations above 30 actual ft³/min as the diaphragm stress becomes too high with the need for greater displacements.

How It Works

The unusual feature of operation of the diaphragm compressor is that of compression of gas by means of a metallic diaphragm which forms an impermeable membrane between t   gas on one side and the hydraulic fluid on the other. The hydraulic fluid is used to pulse the diaphragm which is sandwiched between two metallic plates, firmly held together by flange bolts. See Fig. 21 for a schematic representation of the compressor. The plates have a concave surface facing toward the diaphragm which, in the downstroke position, is essentially flat between the two concavities.

The outer periphery of the diaphragm acts as a sealing member, preventing leakage of gas or hydraulic fluid. The upper plate contains the inlet and discharge valves for the gas, while the lower plate contains the holes through which oil or other hydraulic fluid flows.

In operation, gas flows into the cavity between the lower surface of the upper plate and the diaphragm on the downstroke of the compressor, and is expelled from this volume on the upstroke as the diaphragm is pushed tightly against the underside of the upper head plate by hydraulic fluid beneath it.

The chamber beneath the diaphragm is filled with the hydraulic fluid used to pulse the diaphragm. This fluid is pressurized by a piston moving up and down in the chamber. This piston is driven through a crankshaft by a running gear similar to that used in conventional compressors.

A small piston pump, called a compensator pump, driven by an eccentric bearing on the end of the crankshaft, is provided to supply a small quantity of hydraulic fluid into the chamber beneath the diaphragms on each piston

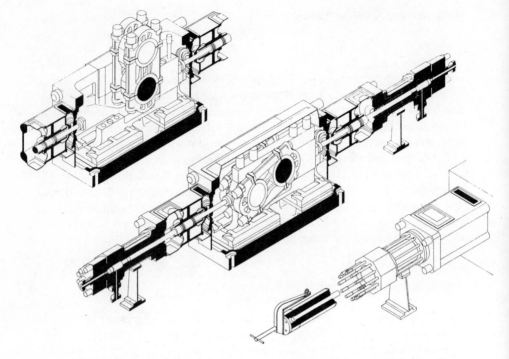

FIG. 20. Large mechanically driven compressor of very high pressures (sectioned).

downstroke. This small quantity of fluid is provided to compensate for compressibility and for leakage past the piston rings.

Other Hydraulic Fluid

The standard type of diaphragm compressor uses oil as the hydraulic fluid. However, diaphragm compressors are available using soapy water as the hydraulic fluid. These compressors are suitable for the compression of oxygen, where oil would be dangerous.

Compressor Design

The diaphragm compressor, though simple in concept and easy to maintain, presents design complexities of a considerable magnitude. For this reason, these machines are not as common as their usefulness might lead them to be.

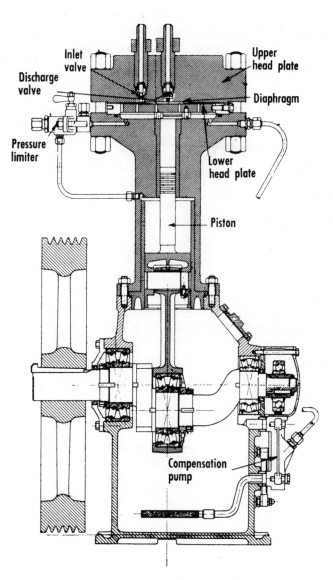

FIG. 21. Diaphragm-type compressor.

The major design complexities are the designs of the upper and lower headplates. If the concavity and the grooves through which gas flows to the discharge valve on the upper headplate are not correctly designed, diaphragms will be improperly and unevenly stressed and will rupture rapidly. If the oil holes are not properly sized and spaced in the lower plate, oil will flow through this plate either too rapidly, again causing uneven stress in the diaphragms and rapid failure, or too slowly, causing inefficient operation.

Unfortunately, there are no simple mathematical formulas for the design of these components since mathematical analysis leads to equations which cannot be solved. The only way to design these machines is through lengthy trial-and-error tests to determine proper design parameters, and even here, design parameters once established for a given machine cannot be transferred to machines of other sizes and pressure ratings.

Advantages and Disadvantages

The advantages of this type of compressor include zero leakage, oil-free compression, no gas seal or packing, and relatively low cost. This compressor is ideal for highly corrosive or highly toxic gases or for an application where absolute purity is necessary.

Disadvantages are low efficiency, diaphragm life generally considered to be 2000 h, and maintenance equal to that of nonlubricated reciprocating compressors.

Manufacturers include Corblin (French), Hofer (German), Lapp Pulsafeeder Co., Leroy, New York, and Pressure Products Co., Hatboro, Pennsylvania.

Rotary Compressors

RALPH JAMES, Jr.

Rotary Screw (Dry-Type)

General

The rotary screw compressor is a positive displacement type machine in which compression is carried out by the intermeshing of two helically formed rotors. The rotors are designated as the male rotor and female rotor. The male rotor has convex lobes while the female rotor has concave flutes.

The gas to be compressed is admitted through a properly located inlet port and completely fills one of the flutes in the female rotor. The discharge end of the rotors are sealed by the compressor end plate. As the rotors rotate, the male rotor lobe enters the pocket in the female rotor, decreasing its volume and compressing the trapped gas. When the desired amount of compression is achieved, the discharge port is uncovered and the compressed gas is expelled. The process is repeated for each successive interlobal space.

There are four lobes on the male rotor and six flutes on the female rotor. The rotors are kept from contacting by timing gears. This arrangement results in six compression cycles per revolution of the female rotor, or four compression

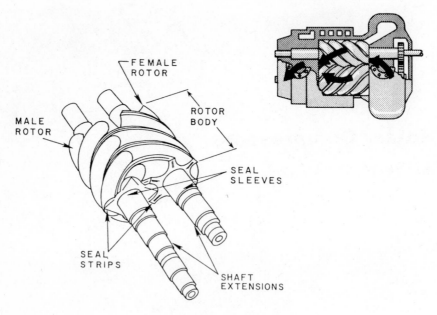

FIG. 1.   Helical compressor rotors.

cycles per revolution of the male rotor. (The female rotor operates at two-thirds of the male rotor speed.)

The amount of compression which takes place in the compressor is determined by the rotor length, helix angle of the lobes, and shape of the compression ratio for a given compressor.

Advantage

The rotary screw compressor is most suited to flows below centrifugal compressor capability (see Fig. 2). Discharge pressures range from a few torr to about 500 lb/in.^{2} gauge. Although centrifugal machines can be applied in portions of this area, efficiencies are usually lower than for the screw machine.

Other features of the rotary screw machine make it an attractive choice for some applications. Since the machine is a positive displacement type, i.e., a definite quantity of fluid is displaced in each revolution of the rotor, a given compressor is not affected by variations in gas molecular weight as compared to the centrifugal compressor. Also, economic comparisons indicate that the

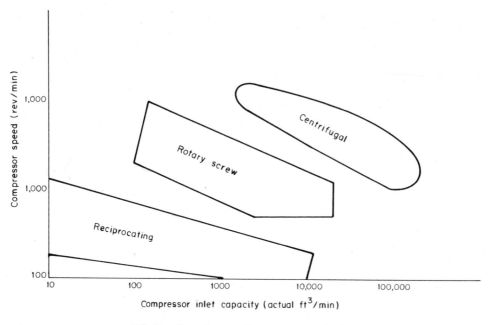

FIG. 2.   General areas of compressor application.

rotary screw compressor will cost less than centrifugals in machine sizes between 200 and 1500 hp.

Although the reciprocating compressor operates with high efficiency, the machine becomes large and expensive in the flow range of the rotary screw machine. The unbalanced forces which result from reciprocating motion are not present in a rotary screw compressor, hence heavy foundations are not required and lower installation costs are realized. Space is conserved with a rotary screw machine because of its small size in comparison to a reciprocating compressor.

When oil-free compression is required, the rotary screw compressor has an advantage since lubricant does not enter the compression chamber. Non-lubricated reciprocating compressors are available, but these are usually low capacity, low horsepower units. Also, rapid wear of nonlubricated parts has severely limited run lengths.

One of the major advantages of the screw machine over the reciprocating compressor is that the former has been successfully applied in handling sticky and polymer-forming gases, and gases containing significant amounts of entrained liquid. In fact, soft deposits on the rotors increase compressor efficiency because clearances between rotors and sealed and back leakage is limited. In a fouling gas service, reciprocating compressors require frequent shutdowns for servicing valves and piston rings. The only caution for a rotary screw machine in a fouling service is that provision be made to flush the compressor prior to or immediately after shutdown to prevent the rotors from sticking. If hard deposits are expected, or if the rotor temperature is high enough to melt polymer particles, continuous flushing is necessary. A hard deposit or a molten polymer will cause thrust bearing overload. White-oil can be injected into the suction line of screw compressors to prevent hard polymer deposits. Cooling water can also be injected to reduce discharge temperatures, flush out solids, or prevent polymerization. The preferred configuration for liquid injection is with the inlet on the top and the discharge on the bottom.

Some Limitations

Perhaps one of the major disadvantages of the rotary screw compressor is that the machine is inherently noisy. All installations require suction and discharge silencers. In some cases, where silencing is critical, an acoustic hood must be fitted.

Control flexibility is limited to speed or recycle control. (An oil-flooded screw compressor is available with a slide valve that moves parallel to the compressor axis to control capacity. This machine is primarily used in refrigeration service.) If a variable speed driver is not applied, excess capacity must be recycled through a cooler for reduced capacity requirements with associated power waste. Suction throttling may be employed, but the resulting increased pressure ratio and discharge temperature limits the amount of throttling which can be done without exceeding machine mechanical limitations.

Because of the close clearances between rotors and between rotors and casing, temperature limits must be carefully observed. Methods for maintaining temperature stability are discussed in the next section.

The rotary screw compressor is sensitive to rotor corrosion and erosion which affects clearances. Any increase in clearances will quickly result in performance deterioration.

One further limitation of rotary screw compressors is that discharge pressures are limited to a maximum of 500 lb/in.^{2} gauge with present casing designs.

Casing Problems Due to Thermal Distortion

Two casing designs are available for rotary screw compressors which are dependent on the size of the machine. These are designated as vertically split and horizontally split. The former is used for normal inlet capacities to approximately 7000 inlet ft^3/min, while the latter is used for capacities above 7000 inlet ft^3/min.

The *vertically split design* consists of two main portions, the inlet casing and the pressure casing. The inlet casing contains the suction nozzle and houses the suction end seal and bearing cartridges. The inlet casing is bolted to the pressure casing which contains the rotors and houses the discharge end bearing and seal cartridges and the timing gears. Bolted to each casing are end covers which, when removed, expose bearings, seals, and timing gears for service. To remove rotors, the casing is opened along the vertical split and the rotors are pulled out in an axial direction.

The *horizontally split design* also consists of two portions. The casing halves are bolted in a horizontal plane. The top half is removed to permit lifting out of rotors. Seals and bearings as well as timing gears are fitted in the casing, and end covers are provided for service to bearings, seals, and timing gears as in the vertically split design.

All casings are water jacketed to maintain dimensional stability, which is important because of the close clearances between rotors and casing. It is extremely critical that the casing cooling system be maintained in good operating condition and that water temperatures be kept within vendor recommended ranges. There have been instances where either plugging of the cooling system or improper water temperature has resulted in differential thermal growth accompanied by rotor casing contact. The consequence is severe rotor and casing damage which could necessitate the replacement of both rotors and casing. In general, these critical cooling requirements favor a fresh water system; salt water cooling is not recommended.

Because of the close clearances, usually 0.001 to 0.015 in. (0.006 in. on C-4101) between rotors and between the rotors and casing, depending on the size and type of machine, it is essential that rotor dimensional stability be maintained. For applications with discharge temperatures above approximately 350°F, a hollow chamber is bored through the center of the rotors to allow the passage of coolant. The coolant used is normally lubricating oil. The upper limit for oil-cooled rotors is about 450°F. For higher discharge temperatures, multicasing arrangements with an intercooler must be used.

Alarms and Protective Devices

Since thermal expansion which affects clearances in the compressor can result in severe wrecks, special consideration should be given to high discharge temperature alarms and shutdowns, as well as to instrumentation which monitors the casing cooling water system. The devices used for temperature monitoring must be set to positively protect the machine from damage should any deviation from the allowable temperature range occur.

Other alarms and sensing devices should be provided to monitor lube oil pressure, bearing temperatures, cooling system, and any other areas necessary for safe compressor operation. As with any installation, the systems must be as reliable as possible.

Mechanical Design

Investigation has shown that in the design of a rotary screw compressor, the components do not differ between suppliers to any great extent. Possible component difficulties are discussed and methods of preventing these difficulties are indicated below.

Rotor

The current shape of the lobes and the male rotor is the circular arc cross section, and the corresponding flutes on the female rotor also have circular arc cross sections. Early rotary screw compressors differ from this configuration in that the trailing edge of the male rotor lobe and leading edge of the female rotor lobe were of involute form, while the leading edge of the male rotor lobe and trailing edge of the female rotor lobe were of circular arc cross section.

The new shape was developed because it is easier to machine and gauge. Sealing strips are easily fitted along the crests of the rotors to prevent back leakage of gas, which is one of the major efficiency debits in the rotary screw machine. Rotor tip speeds are normally 150 to 350 ft/s.

To provide channels for oil cooling, the rotors must be cut prior to final machining, the oil channels formed, and the pieces welded together. There are three common methods used for fabricating this type of rotor. The first consists of inserting and welding stub shafts on either end of the rotor after the oil channel has been bored; the second requires only one stub shaft which is fitted after boring; and the third involves cutting the rotor in the center through the lobe portion prior to machining the lobes, forming the oil cavity, and welding the halves together. The choice of method is normally dependent on the supplier. Whatever the method employed, the weld should be checked thoroughly to determine acceptance of the joints. There have been instances of weld failure in some machines due to inferior fabrication of oil-cooled rotors.

Bearings

Heavy duty rotary screw compressors are equipped with pressure lubricated sleeve-type bearings and tilting pad thrust bearings. Ball and roller bearings are used on light duty machines.

The major components of load on the radial bearings are the gas load, which increases from the suction to discharge of the machine as a function of pressure buildup, and the rotor weight. Maximum bearing pressures are normally 300 to 400 lb/in.^{2}. To facilitate maintenance, the babbitt-lined bearings are pressed into steel cartridges which are easily removed. Since the discharge side of the machine sees the highest pressure, the discharge radial bearings are of slightly greater area to absorb the heavier load.

The thrust bearings are provided to absorb the thrust generated by pressure forces. Most vendors utilize Kingsbury-type tilting pad thrust bearings.

Bearing problems have not been frequent on rotary screw compressors. The bearing chamber is separated from the compression chamber to prevent lubricating oil from entering the gas stream.

Seals

Screw compressor seals are of three basic types: carbon ring seals, labyrinth seals, and mechanical contact seals. Variations in these types are made according to the requirements of the service.

Sealing systems have probably been the most troublesome component in rotary screw compressor application to process services. In general, poor choice of seal materials in addition to arduous operating conditions have resulted in rapid seal deterioration and frequent seal failures in some process services. Therefore, when a rotary screw compressor is evaluated, special attention should be paid to seal selection and vendor experience in the area of seal application to be certain that the proposed system is suitable for process requirements and is not an extrapolation of vendor experience.

Carbon Ring Seals.   These are the seals most frequently applied to rotary screw compressors. This seal permits the seal area to be short in length to allow effective sealing without impairing rotor rigidity.

To preclude the possibility of lube oil entering the decompression chamber,

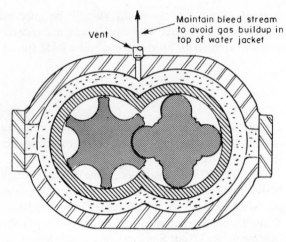

Vent

Maintain bleed stream
to avoid gas buildup in
top of water jacket

Note : Vent connection
usually is located on
bottom of compressor cases.

FIG. 3

a vent area is provided between seal and bearing area. It is common practice to fit a screw thread oil seal between the bearing and vent chambers to prevent leakage of lube oil to the vent and possibly into the compressor. It is important that the vent be connected in such a manner that a differential pressure across the screw thread oil seal cannot exist.

Timing Gears

Perhaps the most critical components in a rotary screw compressor are the timing gears. Their function is to synchronize the rotors and maintain close running clearance between them. The gears must be of high quality to prevent rotor contact or excitation of torsional vibrations.

The timing gears transmit aproximately 10% of the driver input power from the driven male rotor to the female rotor. The remaining power is transmitted by pressure forces imposed by the gas. Therefore, the gears function primarily as synchronizing gears, not as load transmitting gears. The usual procedure is to design the gears to be capable of transmitting 20% of the total power. Since there is no record of timing gear failures, this procedure seems adequate.

Positive retention of timing gears is important. The normal technique of fitting the timing gears is to have the male rotor timing gear shrunk and keyed to the shaft and the female rotor timing gear bolted to a hub which is shrunk and keyed to the shaft. Some interference is provided between the female timing gear and shaft, but experience has shown that this interference and bolting is not sufficient to hold the female timing gear in place. The female timing gear is fitted in this manner to allow it to be moved for required timing adjustment. Positive retention can be assured by providing tapered dowel pins to positively lock the gear after timing is completed. Many vendors provide such pins, but some do not. Therefore, this detail should be checked when evaluating a machine.

Vane-Type Rotary Compressors

Operating Principle

The sliding vane type of rotary compressor consists of a cylindrical rotor in an eccentric casing. Thin, flat vanes (8 to 30) located in slots along radii of the rotor slide in and out as the rotor turns. Centrifugal force drives the vanes outward; contact with the eccentric casing inner surface drives the vanes back inward as rotation gradually decreases the rotor center to cylinder i.d. radius. Rotation within the eccentric casing gradually reduces the volume in chambers trapped by the sliding vanes, thus raising the pressure. See Fig. 4.

This compressor has a rather narrow range of capacity and pressure compared to reciprocating designs because of inherent limits imposed by vane length, rubbing speed on the cylinder wall, and the bending forces acting on the vane when in an extended position. Inlet volume flow rate up to 3200 actual ft^3/min is feasible at discharge pressures from 2 to 129 lb/in.^{2}gauge.

When running at design pressure, the theoretical indicator card is identical to the reciprocator. There is one difference of importance, however. The reciprocating unit has spring-loaded valves that open automatically on small pressure differentials between the outside and inside of the cylinder. The discharge valve, therefore, opens as soon as the discharge pressure is reached and the inlet as soon as suction pressure is reached, even though there may be some variation in the discharge pressure from time to time.

The sliding-vane machine, however, has no valves. The times in the cycle when the inlet and discharge open are determined by the location of ports over which the vanes pass. The inlet porting is normally wide and is designed to admit gas up to the point when the pocket between two vanes is the largest. It is closed when the following vane of each pocket passes the edge of the inlet port.

The pocket volume decreases as the rotor turns and the gas is compressed. Compression continues until the discharge port is uncovered by the leading vane of each pocket. This point must be preset or built-in when the unit is manufactured. Thus the compressor always compresses the gas to design pressure, regardless of the pressure in the receiver into which it is discharging

ROTOR WITH NON-METALLIC
SLIDING VANES.

GAS IS GRADUALLY COMPRESSED
AS POCKETS GET SMALLER.

DISCHARGE

AS ROTOR TURNS, GAS IS TRAP-
IN POCKETS FORMED BY VANES.

COMPRESSED GAS IS PUSHED OUT
THROUGH DISCHARGE PORT.

FIG. 4. The steps in compression for a sliding-vane rotary compressor.

(see Fig. 5). This results in slightly more power at off-design pressures as compared to the ideal reciprocating P–V card.

Advantages Compared to Other Rotary Types

1. Low cost
2. Highest compression efficiency and overall efficiency of all rotary types, primarily because of very low slip
3. Very high volumetric efficiency, hence high flow rate with respect to machine size
4. Low starting torque requirement

Disadvantages

1. Continuous lubrication of internal rubbing parts is required (10 times as

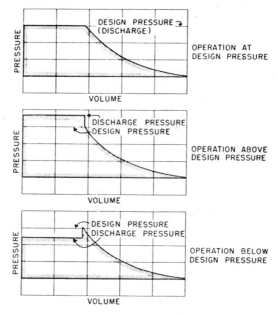

FIG. 5. Types of theoretical indicator cards obtained by any rotary compressor having built-in (fixed) porting.

much as required for a reciprocating compressor)
2. Maintenance is relatively high due to wear of rubbing parts and use of antifriction bearings, necessitating frequent mechanical inspection
3. Sensitive to solids in gas stream, necessitating filters for air service

Construction

The machine is built in two types. One, used exclusively for stationary service, is fitted for standard lubrication from a force-feed lubricator, is water jacketed, and uses water-cooled intercoolers on multistage units.

The other has a combination lubrication and cooling system consisting of a flood of oil to the machine at all times. The heat of compression is picked up by the oil and dissipated through either a radiator-type oil cooler to the atmosphere or a water-cooled exchanger. Oil is separated from the air before being cooled.

Liquid-Ring Compressor

Applications

The liquid-ring compressor is a unique type of rotary machine which utilizes a liquid annulus held by centrifugal force around the inside of a casing by a single impeller to compress vapor (see Fig. 6). It has the low cost per unit volume capacity characteristic of the rotary class and is thus competitive for vacuum and other low-pressure air and gas services. Since it operates without internal oil lubrication, it is applicable where lube oil contamination of the process gas is undesirable or unacceptable.

It is ideally suited to handling gases sensitive to temperature rise, such as acetylene, because the liquid compressant provides continuous contact cooling, allowing the compression to follow a nearly isothermal path. The discharge

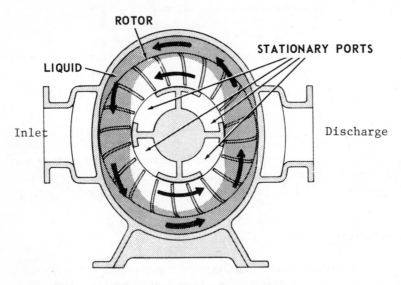

FIG. 6. The liquid-piston compressor.

temperature is typically only 10 to 15°F higher than the compressant liquid inlet temperature. The effectiveness of the cooling also permits high pressure ratios without intercooling, e.g., single-stage atmospheric air compressors discharging to 110 lb/in.^{2}gauge. Inlet volume flow rates up to about 20,000 actual

ft^3/min at 15 to 200 lb/in.^{2}gauge are feasible at lower actual ft^3/min, say about 2000 actual ft^3/min pressures to about 100 lb/in.^{2}gauge are possible.

The liquid ring compressor is capable of performing other unit operations simultaneously with compression because of the internal liquid circulation principle. Dust can be scrubbed from the gas stream and continuously removed by the compressant liquid. Vapor components can be selectively absorbed into a properly selected solvent compressant liquid. Liquid carryover can be removed by absorption into the liquid. Hot inlet gas can be cooled as it is compressed.

Corrosive gases can be handled by means of a compatible compressant liquid. For example, the liquid ring compressor is widely used for dry chlorine vapor (tank car evacuation, etc.) with concentrated sulfuric acid as the compressant.

Advantages of Liquid Ring Compressors

The special advantages of liquid ring compressors over other compressor types may be summarized as follows:

1. Low discharge temperature capability useful for polymerizing gases such as acetylene and for gas mixtures with exothermic reactions
2. Selection and processing of compressant liquid allows various unit operations simultaneous with compression: absorption, scrubbing, and cooling
3. Insensitive to liquid carryover, if compatible with compressant liquid
4. Insensitive to solid fines carryover
5. Oil-free operation
6. Availability potential is high and maintenance costs are low because the machine is extremely simple, with only one moving part
7. Maintenance in very corrosive services is lower than for other types of compressors
8. Roughly 5 times as efficient as ejectors in vacuum service
9. Inlet temperature can be high if the compressant liquid is selected for compatible vapor pressure characteristics
10. Low discharge pressure pulsations, not requiring dampeners

Disadvantages of Liquid Ring Compressors

1. Low overall efficiency (35 to 50%) due to power expended in circulation of

the compressant liquid around the machine and back and forth within the
compression compartments

2. High compressant liquid flow requirement due to low overall machine
   efficiency
3. Compressant liquids other than water may require specially engineered
   separation facilities as well as cooling in recirculating systems

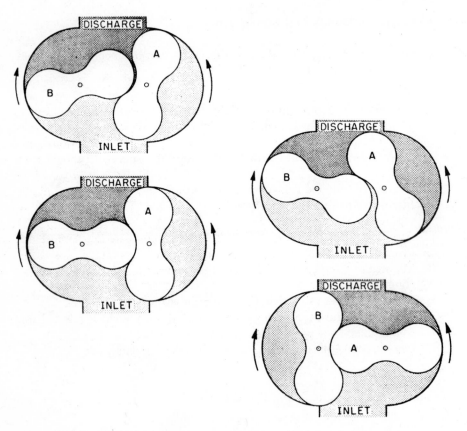

FIG. 7. The operating cycle of a two impeller straight-lobe rotary compressor.

Straight Lobe Blowers

One of the oldest and simplest types of rotary compressors is the "Roots" or straight lobe "blower." This type employs two identical ductile iron rotors, each with two rounded lobes, giving the rotors a figure 8 shape in cross section. The rotor lobes intermesh and are held apart by timing gears. They are straight with respect to the shaft axis, like rounded teeth on a spur gear. As the rotors turn, gas is swept between the rotor and the casing wall toward the discharge with no volume reduction. Compression takes place as the discharge port is uncovered. Gas from the discharge line then flows backwards into the casing until the pressure in the cavity reaches discharge line pressure. Further rotor turning sweeps the mixture back into the discharge line. See Fig. 7.

Most straight lobe types have two lobes on each rotor, but some models use more lobes.

Models are commercially manufactured from fractional horsepower to 3000 hp. Peak efficiency occurs at 5 to 7 $lb/in.^2$ gauge.

This compressor type has both low cost and low maintenance requirements for the low-pressure services to which it is applicable. It is less efficient than screw compressors because of slip and the high frequency flow direction reversal which takes place at the discharge port. Some models have considerable tolerance for liquid, but solids deteriorate the critical internal clearances which limit slip.

Straight lobe blowers are used as displacement meters in pipelines, as boosters in low-pressure pipe lines for conveying powders and pellets in gas well gathering systems, as engine superchargers, and in desalinization and other evaporative processes where large volumes of vapor are removed at high vacuums.

Thermal Compressors

RALPH JAMES, Jr.

General

A thermal compressor is an ejector which operates with suction pressure above atmospheric—usually with a compression ratio of less than 2 (subcritical flow).

Usually an ejector is thought of as a compression device operating with an inlet pressure considerably below atmospheric and discharging at a level which, as a maximum, would slightly exceed atmospheric pressure. Under these conditions, while actually a compressor, an injector is classed as a vacuum producer or "pump."

Ejectors are devices for raising the pressure of liquids or vapors which operate by entrainment of a pumped fluid into a high velocity jet of a higher pressure motive fluid stream. (See Fig. 1 for a cross-section view of an ejector.) Ejectors have no moving parts but are much less efficient than mechanical pumps and compressors, and therefore are only applied where there are large quantities of low-pressure motive steam or compressed gas available at low cost. Because they can handle the large volume flow rates at the low pressures required, they are commonly used on vacuum distillation towers and surface-

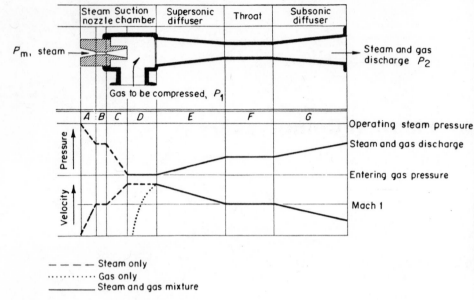

FIG. 1. Pressure and velocity variations with a steam jet ejector handling gas (critical flow). *A*: Subsonic steam velocity generated to Mach 1 in a converging nozzle as steam pressure drops. *B*: Stabilization with pressure constant, velocity constant at Mach 1. *C*: Supersonic steam velocity raised in diverging nozzle as pressure drops. *D*: Since suction chamber is at the lowest pressure in the system, the air flows into the chamber and is entrained in the steam jet. *E*: Supersonic mixture pressure is increased in converging diffuser until velocity drops to Mach 1. *F*: Stabilization with pressure constant, velocity constant at Mach 1. *G*: Subsonic mixture pressure is increased in diverging diffuser as velocity drops.

type steam condensers to compress vapors which are not condensible at available cooling water temperatures to pressures at which they can be condensed or vented from the vacuum system.

Definitions

Thermal Compressor (thermocompressor)—An ejector which operates with suction pressure above atmospheric and usually with a compression ratio of less than 2 (subcritical flow).

Boosters—Ejectors applied for the high volume, low pressure stages of a multistage system, up to the pressure level where condensers are effective with available cooling water temperature.

Pressure Ratio—The ratio of the mixture discharge pressure, P_2, to the entrained fluid inlet pressure, P_1.

Expansion Ratio—The ratio of the motive fluid inlet pressure, P_m to the entrained fluid inlet pressure, P_1.

Entrained Fluid—The service fluid which the ejector compresses. This term is preferred over "pumped" fluid in ejector engineering.

Characteristics

Compression ratios may be greater or less than 2 to 1, but for most applications will be less than this critical ratio. This changes the velocity pattern in the diagram of Fig. 1 in that throat velocity in Area F will be below sonic. Ejector characteristics will not be the same as for an ejector with sonic diffuser throat velocity.

Figure 2 shows typical performance curves on both types of ejectors. In each, the left-hand curve, or group of curves, shows variation of intake pressure with capacity. The right-hand curve represents discharge pressure. In the vacuum ejector, it is the *maximum* discharge pressure obtainable. The *actual* discharge pressure may be lower with no change in the suction pressure–capacity curve. This is not so with thermal compressors. Any change in actual discharge pressure is reflected in a similar change in suction pressure for the same capacity in lb/h.

In each, there are four possible variables: steam weight flow or steam pressure on a fixed nozzle, intake pressure, discharge pressure, and capacity in lb/h. Table 1 is based on changing one of these at a time. The influence on the others is noted. (*Ratio* means compression ratio.)

Two characteristics stand out in the functioning of a thermal compressor. One is the effect of varying steam weight flow for capacity control. The other characteristic is the way a change in discharge pressure causes the intake pressure–capacity curve to move or *float*. This is at an almost constant compression ratio.

These characteristics are only true within reasonable areas. For example, increasing steam weight flow on a thermal compressor too far could possibly increase throat velocity to the sonic range; the ejector would then act like the normal vacuum ejector.

Ejectors may be arranged in either parallel or series. When two or more steam ejectors are placed in series to form a multistage arrangement, it is usual,

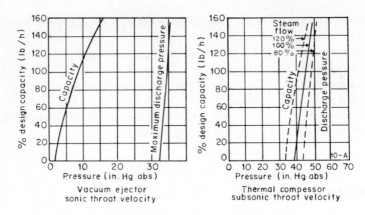

FIG. 2. Comparison of the operating characteristics of a vacuum ejector and a thermal compressor.

TABLE 1

Vacuum Ejector	Thermal Compressor
Increased Steam Flow	
Increases *maximum* discharge pressure; practically no other change	Reduces intake pressure. Increased ratio. Alternately, can operate at constant ratio and increased capacity
Increased Intake Pressure	
Reduces ratio; increases capacity	Reduces ratio; increases capacity
Increased Discharge Pressure	
No change (until reaches breaking pressure)	Entire family of curves moves with the discharge. Ratio remains practically constant
Increased Capacity	
Reduces ratio; increases intake pressure	Reduces ratio; increases intake pressure

if water temperature is sufficiently low, to interpose a condenser between successive elements to condense the steam used by the prior jet plus any condensable vapor in the gas being compressed. This materially reduces the steam required for the next stage since the weight of mixture remaining to be handled is much less. Both barometric and surface intercondensers may be used.

Incentives for Applying Ejectors

Ejectors are attractive alternatives in certain compression or vacuum services for the following reasons:

Low equipment cost
Low installation costs
Higher reliability in severe services
Tolerance for entrained liquids, even in slug form, intermittently
Corrosion damage is easily repaired and at relatively low cost
No stuffing box sealing required
Simple operation, no moving parts

The principal disadvantage of ejectors is very low efficiency (1 to 20%) when compared to mechanical compressors. Therefore they are usually an economic choice only if there is low cost, low pressure steam or compressed gas available.

Ejector Ratings

Ejectors are rated on the weight flow of gas compressed rather than on volume flow as is usual with other compressor types. Ratings are expressed in lb/h of the gas or gas mixture handled. Steam (or other motive medium) is also stated in lb/h.

Normally, manufacturers' basic ratings are on the weight rate of "equivalent air," which is air at 14.696 lb/in.^{2}abs and 70°F and containing normal atmospheric moisture. The user must specify the desired capacity in such terms that equivalent air can be calculated and the best ejector can be selected.

Capacities desired are preferably specified in one of three ways:

1. Lb/h of a saturated air–water vapor mixture
2. Lb/h of a superheated air–water vapor mixture

3. Lb/h of a mixture of gases either including or excluding water vapor

The proportion of air to vapor in Method 1 is known and no further information is needed for conversion to equivalent air.

In Method 2 it is mandatory that the relative weights of air and vapor be specified, otherwise it is impossible to determine the molecular weight and equivalent air.

Method 3 requires that a complete analysis of the gas be given or that the user specify the average molecular weight. The analysis may be on a volumetric (molar) or a weight ratio basis.

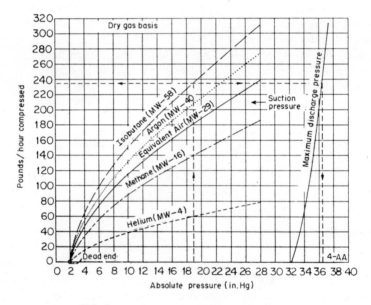

FIG. 3. Typical ejector performance curves.

Transfer from specified lb/h to equivalent air is based on tests run under the auspices of the Heat Exchange Institute [1], and are accepted and used by all ejector manufacturers.

Figure 3 is a typical performance curve of a single-stage ejector showing what happens to ejector performance with changes in the characteristics of the gas being handled. In this example, motive steam flow is held constant. It can be seen that for every ejector there is a definite relation between absolute suction pressure and capacity. The right-hand curve shows the maximum stable operating discharge pressure obtainable with this particular ejector with a specified steam pressure and varying capacities.

Figures 2 and 3 with relevant descriptions were supplied courtesy of the Ingersoll-Rand Company, Phillipsburg, New Jersey.

Reference

1. Heat Exchange Institute, *Standards for Steam Jet Ejectors*, 3rd ed., New York, 1956.

Turboexpanders

JUDSON S. SWEARINGTON

Turboexpanders have been in wide use since about 1950 for the separation of gases by partial condensation, and the number of applications is rapidly increasing.

The isentropic expansion process (in combination with compression) is an efficient refrigeration process for low temperature, and the equipment is of reasonable cost.

It is all but universally used for air separation and for hydrogen and helium liquefaction, and is widely used for the recovery of ethane and heavier compounds from natural gas. Other uses are in the recovery of ethylene, the processing of hydrocarbon gases, demethanization, natural gas dew-point adjustment, and natural gas liquefaction.

The turboexpanders for such applications are required to be highly efficient, to have excellent reliability, and to have other qualities to be discussed. As a consequence, they are also useful for high-speed power drives, for power recovery, and for power generation.

Accordingly, this article is directed to the following subjects:

Design Criteria for Expansion Processes

The turboexpander in combination with a compressor and heat exchanger may be regarded as a refrigeration machine (see Fig. 1). The heat is shown absorbed by the expander discharge stream and rejected in the compressor aftercooler.

The second law of thermodynamics efficiency is best illustrated by regarding the compressor as isothermal, based on the heat rejection average temperature (T_2), and the expander likewise isothermal at the heat absorption average temperature (T_1).

Based on 1 Btu of cooling absorbed at T_1, the available enthalpy expressed in Btu's in the gas to the expander must be $(1/E_e) \times 1.1$. The factor 1.1 is to allow for the heat exchanger temperature difference $(T_{1h} - T_{1e})$. Further, the work required by the compressor is the work quantity multiplied by the reciprocal of the compressor efficiency. Thus the compressor requires work (also expressed in Btu's) equal to

$$\frac{1}{E_c} \times \frac{T_2}{T_1} \times \frac{1}{E_e} \times 1.1$$

The expander delivers work equivalent to the cooling at T_1 multiplied by about 1.1 (to allow for the additional ΔT consumed by the heat exchanger temperature difference), so after crediting this expander work to the power input the net work is

$$W = \frac{T_2}{T_1 E_c E_e} \times 1.1 - 1.1$$

For $E_c = 0.7$ and $E_e = 0.85$,*

$$W = 1.1\left(\frac{T_2}{T_1} \times \frac{1}{0.7} \times \frac{1}{0.85}\right) - 1.1$$

$$\text{Reversible work} = \frac{T_2 - T_1}{T_1}$$

so the second law efficiency of the system is equal to

$$\frac{T_2 - T_1}{T_1}\left/\left(\frac{T_2}{T_1} \times \frac{1}{0.7} \times \frac{1}{0.85} \times 1.1 - 1.1\right)\right.$$

which is shown plotted against T_1 in Fig. 2.

The second law efficiency is seen from Fig. 2 to improve as the required temperature falls, and the use of expanders for refrigeration is at some power disadvantage above about $-100°F$; however, for expediency, they often are

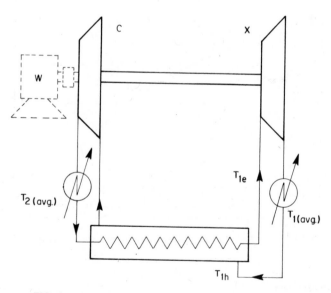

FIG. 1. Expander/compressor as a refrigeration system.

*The refrigeration usually is useful over the full temperature range in which it is available, so the expander efficiency is based on the average heat absorption temperature.

used anyway, especially if the pressure drop is already available. This improvement in second law efficiency at lower temperatures favors expanders for processes that require low temperature. The expander should be located in the system at the point where the heat should be withdrawn, and this is usually at or near where the temperature is lowest. Such location often means that the expanding stream partially condenses as it passes through the expander, and the expander should be capable of operating on a condensing stream without damage or efficiency deterioration.

Since the cost of refrigeration is rather high, the temperature approach in the heat exchanger should be correspondingly close. Economics dictates an average temperature difference of the order of 10°F down to 200°R. Below that, it should drop in proportion to the square root of the absolute temperature. Further, the high cost of refrigeration significantly accents the importance of cycle optimization, heat insulation, and other refrigeration economies.

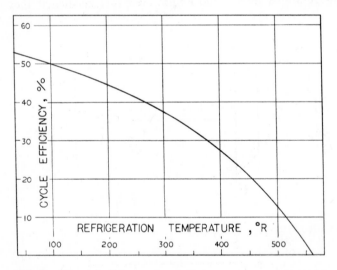

FIG. 2. "Carnot" efficiency of expander/compressor vs temperature.

Air Separation

Steel making has provided the largest recent market for oxygen. Nitrogen is the so-called by-product, but most of it is used (at low or nominal price) for inert gas applications and as a material to be liquefied to serve as the medium for storage and handling of low-temperature refrigeration. The 1% of argon in air likewise has found a large market in shield-arc welding.

Figure 3 shows a typical low-pressure air separation cycle yielding gaseous

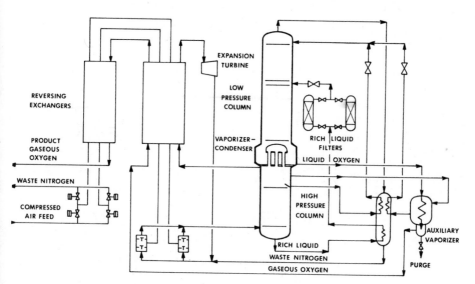

FIG. 3. Low-pressure air separation cycle for gaseous products. (Courtesy Air Liquide.)

products. It is to be noted that the expander is near the coldest point (to attain best second law efficiency, see Fig. 3). Also, if of sufficient capacity, the power from the expander is usefully recovered. Further, refrigeration economy is practiced: the refrigeration required is equal to the enthalpy difference between inlet pressurized air and discharged products, plus heat leak.

The rate of sale or use of oxygen (and nitrogen) often varies, and to meet this variation provision is usually made in air separation plants to produce portions of the product oxygen and nitrogen as liquids. This also provides liquid for markets requiring hauling and storage.

Figure 4 shows a modern air separation cycle for producing liquid as well as gaseous products—and a crude liquid argon stream. In a low-pressure plant, the heat capacity of the pressurized air is little different from that of the products, so the heat exchanger balances reasonably well. In a liquid plant the product is removed cold, resulting in the heat exchanger not being well balanced. The portion of this refrigeration deficit in the upper temperature range can be supplied by Freon or ammonia refrigeration.

At the next lower temperature range in the heat exchanger, use is made of Joule-Thomson cooling plus isentropic expansion at high pressure with a turboexpander (and high-pressure liquefaction and flashing to a lower pressure and temperature) to provide the low-temperature refrigeration.

If only a modest amount of liquid product is required, the cycle typically shown in Fig. 5 may be used. It is really a "low-pressure" cycle with extra

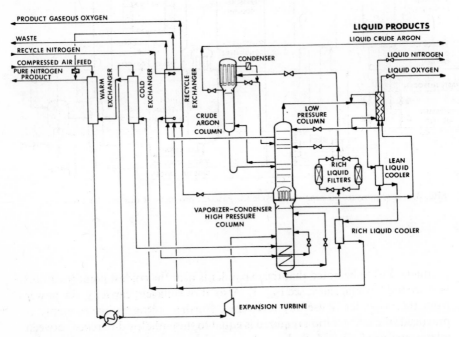

FIG. 4. Process for air separation producing gaseous products along with substantial amounts of liquid. (Courtesy Air Liquide.)

refrigeration made available when needed by significantly raising the inlet air supply pressure and flow.

In another arrangement of the "low-pressure" cycle [1] where compressor pressure is varied to control the extra refrigeration, the towers are at elevated

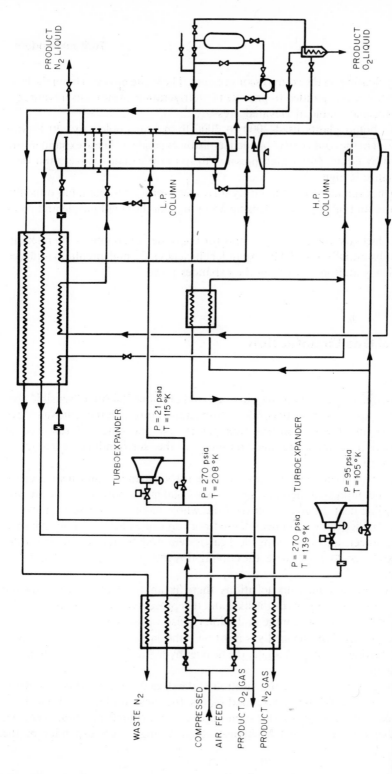

FIG. 5. Air separation process, intermediate pressure plant in which the amount of liquid products is varied by varying the air supply pressure. (Courtesy Linde Division of Union Carbide Corp.)

pressure, floating on the compressor pressure. The waste nitrogen from the low-pressure tower is expanded to provide the refrigeration. Among its advantages are products available at elevated pressure.

Older liquid plants used 2000 to 3000 $lb/in.^2$ air cooled to $-40°F$ with Freon, and in the cold zone expanded in piston expanders. These expanders are damaged by liquid, suffer high pressure-drop in the exhaust valves, and their maintenance is high.

A turboexpander for similar duty solves these problems, but it has difficulty with water and oil vapor condensing and plugging the small nozzle and rotor passages.

The chief cost for air separation is for the power to compress the inlet air. This is substantially affected by expander efficiency, so a high premium is placed on efficiency and on recovering the expander power.

Natural Gas Liquefaction

The liquefaction of methane requires removal of about 180 Btu of sensible heat per pound to cool it down to condensation temperature, and the removal of 220 Btu of latent heat of condensation per pound, the latter all removed at condensation temperature and rejected to the surroundings at ambient temperature.

The simple isentropic expansion (with turboexpanders and heat exchangers) of a compressed methane stream will condense a portion of the stream. The sensible heat difference between the whole incoming pressurized stream and the smaller returning uncondensed portion causes heat exchanger imbalance. In effect, all of the sensible heat is removed by the low-temperature turboexpander, so, powerwise, the cooling-down step in the process is not very efficient.

However, in numerous locations there is an available "free" pipeline pressure drop at city gates or the like, and this process is used to condense 10 to 20% of the stream for storage and winter season peak-shaving.

To avoid trouble, the feed gas must be sufficiently cleaned of water vapor, carbon dioxide, and dust. Usually the heavier hydrocarbons can be left to blend with the liquid.

A process for the total condensation of the stream using a turboexpander is shown in Fig. 6. For large base-load plants, the cycle in Fig. 6 [2, 3] has been proposed. It utilizes a single step of multiple component refrigeration to correct the imbalance in the heat exchanger, then a single turboexpander in the

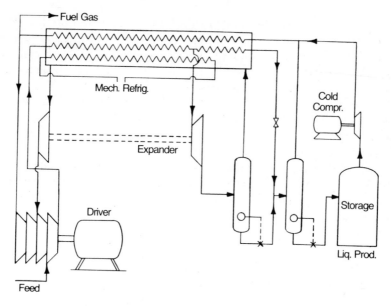

FIG. 6. Turboexpander cycle applied to the full liquefaction of natural gas.

condensation range, and liquid flashing to storage. No plant using this process has been built.

Ethane–Propane Recovery from Natural Gas

This process requires that the feed gas be reconditioned by drying, preferably to a dewpoint near $-100°F$, and the CO_2 content reduced if its concentration is such that it might reach its frost point in the process. If the drying is to no lower than a dewpoint of -50 or $-60°F$, the turboexpander is sure to load up with ice in a day or a few days. If the CO_2 frost point is reached in the turboexpander, frost may not accumulate there but it will probably foul downstream equipment. The low-temperature portion of the process is shown in Fig. 7.

Numerous variations are used for improvement in efficiency such as the use of two or more turboexpanders and intratower heat exchange.

Factors that would reduce overall plant efficiency are: poor uniformity of temperature difference in the main heat exchanger, expansion of the gas at higher than necessary temperature [for the false reason that it produces more power (refrigeration) at the higher temperature], poor refrigeration economy such as excessive heat injection into a tower reboiler, insufficient attention to pressure drop losses, and especially not protecting the turboexpander from damage by solids entrained in the gas.

Turboexpanders are available to expand condensing streams, and this allows the expander to be operated at the low temperature where the process is most efficient.

An important advantage of the expansion process (over cold oil absorption) is the ease with which plant adjustments can be made to attain the required recovery needs which may be changed from time to time. A somewhat higher pressure drop for the plant provides the additional required refrigeration for a higher recovery.

Other Recovery (Purification) Processes

Turboexpanders are used in conventional ways to chill or further cool streams to increase recovery of components such as ethylene or to remove contaminants such as CH_4, N_2, or CO from hydrogen.

Helium is recovered from natural gas by condensing out the hydrocarbons (and then revaporizing the latter).

Substantial amounts of natural gas containing objectionable N_2 content have been found. The heating value may be increased by partially condensing the hydrocarbons and revaporizing them. The uncondensed nitrogen-rich residue gas is expanded to near atmospheric pressure to produce refrigeration and (if it is not fractionated) used as gas turbine fuel.

For extremely low-temperature processes such as hydrogen liquefaction and especially for helium liquefaction it is advantageous to adjust the temperature difference in the main heat exchanger at one or two intermediate temperatures by the use of intermediate expanders.

Other Uses for Turboexpanders

1. Power recovery from streams now being throttled. This use results from the fact that the turboexpanders operate at high efficiency and are available in advanced proven designs.

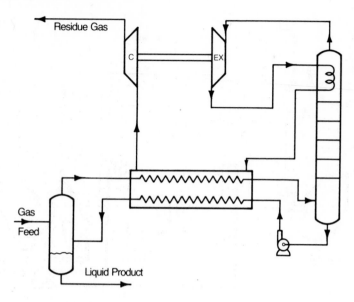

Residue Gas

C   EX

Gas
Feed

Liquid Product

FIG. 7.   Cycle for LPG recovery from natural gas.

2.   Adjustments of natural gas dew point by partial condensation of any heavy constituents.
3.   High-speed drivers.
4.   Units for recovering power from low-grade heat sources based on expanding a low-boiling working fluid.

Turboexpander Description

Figure 8 shows a typical simple compressor-loaded turboexpander. The rotating assembly is seen to comprise a shaft mounted in two oil-lubricated journal/thrust bearings and having the turboexpander rotor on one end and compressor impeller on the other end.

The expander shaft seal is a close-clearance labyrinth (fitting a conical section of the shaft) with provision at an intermediate point for injection of a compatible seal gas stream to prevent leakage of process gas and to prevent intrusion of oil. The compressor end likewise has a shaft labyrinth seal as

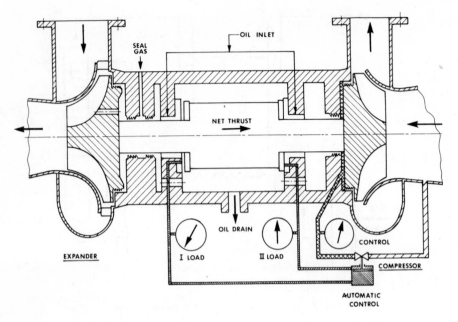

FIG. 8.   Turboexpander cross section. (Courtesy Rotoflow Corp.)

shown. Provision for introduction of seal gas usually is provided (not shown in the figure).

It is desirable to have "high rigidity" journal bearings so the rotor does not go through its oil film critical. At speeds above this critical the assembly tends to rotate about its center of gravity and if unbalanced by an ice deposit it will wipe the seals.

The flow through the expander is controlled by variable area primary nozzles controlled by the pneumatic actuator. This method of control allows high adiabatic efficiency over a wide flow range.

The speed should vary somewhat with inlet pressure to maintain high efficiency, and this is inherently attained. For example, greater power resulting from higher pressure would drive the compressor at a higher speed.

The expander is quite rugged and reliable but should be protected from solids that may be entrained in the process gas such as welding beads and drying bed dust. An 80- or 100-mesh screen will protect it against most such debris, but occasionally it is necessary to filter out fine silica or iron sulfide or Teflon/graphite dust from dry compressor piston rings.

In many applications the incoming gas may be at its dew point or may reach its dew point within the expander. Such resulting fog particles are subjected in the rotor to centrifugal force and to deceleration force amounting to many thousand times the force of gravity. To avoid damage from the resulting particle drift it is necessary that the expander rotor blades be made parallel to this drift direction at all points in the rotor [4, 5].

At pressures above 100 lb/in.^{2} there is a possibility for upsets to overload the thrust bearings. To meet this problem, each thrust bearing is fitted with a thrust load meter [6] consisting of a pressure gauge reading the oil pressure in the thrust bearing oil film as shown in Fig. 8. If the thrust (in either direction) is excessive, an adjustment can be made by varying the balancing seal pressure by means of the thrust control valve which adjusts the back pressure on the balancing seal pressure as it vents to compressor suction [6]. Bearing lubrication is usually at 100 to 150 lb/in.^{2} difference in a closed system as shown—no vent to atmosphere. Such machines often operate continually for over 5 years without inspection or repair.

Other methods of absorbing the expander power are by mechanical drives such as gear and generator or pump or the like. These require conventional shaft seal systems to avoid gas leakage to the atmosphere.

If the power developed is less than 100 hp, it is frequently not worth recovering and is absorbed by an oil turbulator circulating lubricating oil through a cooler.

Turboexpanders are in operation at pressures ranging up to 3000 lb/in.^{2}. Higher pressures are possible, but no such applications have materialized.

Cost

A complete unit usually consists of a large skid on which the turbo-expander/compressor is mounted along with the lube system and control panel.

TABLE 1 Cost for Expander/Compressor Systems with Lube and Instrumentation (no trip valve)

hp	$ (1978)
50	50,000
150	64,000
500	101,000
1,500	150,000
5,000	270,000
10,000	380,000

The lube system is built for high reliability with standby oil pump, dual filters, dual oil coolers, and pressurized oil reservoir. Many have stainless steel piping, and often API specs are requested.

The instrumentation starts the standby pump on low oil pressure. It also has eight or more alarms and eight shutdown circuits to sense low oil pressure, various temperatures, oil levels, overspeed, etc., to actuate and lock-in alarm and shutdown signals. The alarms illuminate signal lights and sound an alarm. The shutdown signals actuate a quick-acting shutdown valve. The first shutdown signal to be actuated will so indicate by blinking. Other adjuncts are vibration sensors, tachometer, and speed control.

The average costs for complete systems are approximately as shown in Table 1.

There are adders for special materials, high pressure, etc.

Safety trip valves and compressor surge control systems are rather expensive.

Symbols

E efficiency
T temperature ($^{\circ}$R)
W driver work (Btu)

Subscripts

c compressor
e expander
h heat exchanger
1   inlet
2   discharge

References

1. U.S. Patent 3,070,966, Superior Air Products, Inc.
2. *Oil and Gas Journal, 40*, 67 (October 6, 1969).
3. U.S. Patent 3,690,114.
4. U.S. Patent 3,610,775.
5. *Chemical Engineering Progress, 68*(7), 95–102 (1972).
6. U.S. Patent 3,895,689.

Calculation Procedures

(see also Design Shortcuts)

RALPH JAMES, Jr.

Part A. Centrifugal Compressors

The calculation procedures described here were developed by the Elliott Company and refer to Elliott Frame sizes. However, other manufacturers' machines can also be estimated from these procedures, provided that data equivalent to Table 2 are available.

The methods outlined are:

1. The N method (so named because of the extensive use of the polytropic exponent n). It is used (a) when the fluid to be compressed closely approximates a "perfect" gas such as air, nitrogen, oxygen, hydrogen; and (b) when a chart of the properties of the gas or gas mixture is not available.
2. The Mollier method involves use of a Mollier diagram and is used whenever a plot of the properties of the fluid being compressed is available. Mollier diagrams, of course, are readily available for most pure gases at "conventional" pressures and temperatures. However, in cryogenic areas or at

very high pressure, gases may behave most peculiarly. Gas properties in these regions heretofore have been estimates arrived at through rather empirical methods.

The same is true of mixtures of gases, yet the preponderance of gas compression problems involve gas mixtures. Fast computers, however, now permit plotting of a Mollier diagram with the only input required being the gas composition and the limiting pressure and temperature values.

Compressor Calculation by the N Method (English Units)

Step 1.   If a *pure* gas, begin with Step 2. If a *mixed* gas, calculate properties as shown in Table 1 (see Table 2).

Step 2.   Calculate inlet flow (Q_1):

$$Q_1 = v_1 \times m$$

$$v_1 = Z_1 RT_1/144p_1$$

$$m = \frac{\text{mol/h} \times M_r}{60} \ (\text{lb/min})$$

$$R = 1544/M_r$$

Assume Z_1 to be 1 or use Fig. 1. If using Fig. 1:

$$p_{R_1} = p_1/p_c, \qquad T_{R_1} = T_1/T_c$$

Step 3.   Select compressor frame. Given inlet volume (Q_1), use Table 3.

Step 4.   Calculate average compressibility (Z_{av}):

$$Z_{av} = \frac{Z_1 + Z_2}{2}$$

Z_1 from Step 2. Z_2 as follows:*

$$T_2 \ (\text{approx.}) = \frac{X}{\eta_{ad}}(T_1) + T_1$$

Z_1 from Step 2. Z_2 as follows:*

$$T_2 \text{ (approx.)} = \frac{X}{\eta_{ad}}(T_1) + T_1$$

From Fig. 2, find X (temperature rise factor) and η_{ad} [using pressure ratio (r) as given, k from Step 1 or Table 2, η_p from Table 3]. Then calculate Z_2 (as in Step 2, using Fig. 1) using p_2 (given) and T_2 as calculated.

Step 5. Calculate polytropic head (H_p).† Using Fig. 3, determine H_p/Z_{av}. Multiply by Z_{av} to obtain head or, for greater accuracy:

$$H_p = \frac{{}^n Z_{av} R T_1}{n - 1}\left[\left(\frac{p_2}{p_1}\right)^{\frac{n-1}{n}} - 1\right]$$

$$\frac{n-1}{n} = \frac{k-1}{k(\eta_p)}$$

η_p from Table 3.

Step 6. Find number of stages required. First, from Fig. 4, find *maximum* permissible head per stage:

$$\text{Stages} = \frac{H_p \text{ (Step 5)}}{\text{maximum head per stage}}$$

If maximum head per stage from Fig. 4 is over 12,000 ft, use 12,000.

Step 7. Find speed required:

$$\text{Speed} = \text{nominal speed} \sqrt{\frac{H_p}{12,000 \times \text{no. of stages}}}$$

Nominal speed from Table 2 (at 10,000 or 12,000 ft head depending on impeller selected).

Step 8. Find shaft horsepower required:

---

*This approximate T_2 may differ slightly from the actual discharge temperature since the effect of compressibility upon temperature rise has not yet been considered.
†Adiabatic head may be calculated by using the 100% efficiency line in Fig. 3.

TABLE 1

Gas Mixture	(1) Mol% Each Gas	(2) Mol/h Each Gas	(3) Molecular Mass (Table 2)	(4) (1) × (3)	(5) Mass %	(6) T_c (Table 2)	(7) p_c (Table 2)	(8) (1) × (6)	(9) (1) × (7)	(10) $C_{p,m}$ (Table 2)	(11) (1) × (10)
……	……	……	……	a	$a/d \times 100$	……	……	……	……	……	……
……	……	……	……	b	$b/d \times 100$	……	……	……	……	……	……
……	……	……	……	c	$c/d \times 100$	……	……	……	……	……	……
				d (Apparent molecular mass of mixture)				$T_{c\,(mix)}$	$p_{c\,(mix)}$		$C_{p,m\,(mix)}$

Calculate $k_{(mixture)} = \dfrac{C_{p,m\,(mix)}}{C_{p,m\,(mix)} - 1.99}$

TABLE 2 Gas Properties (most values taken from *Natural Gas Processors Suppliers Association Engineering Data Book—1972*, 9th edition)

Gas or Vapor	Hydrocarbon Reference Symbols	Chemical Formula	Molecular Mass	Specific Heat Ratio $k = c_p/c_v$ at 60°F	Critical Conditions Absolute Pressure p_c (lb/in.²abs)	Critical Conditions Absolute Temperature T_c (°R)	$C_{p,m}$[a] At 50°F	$C_{p,m}$[a] At 300°F
Acetylene	C_2=	C_2H_2	26.04	1.24	905	557	10.22	12.21
Air		$N_2 + O_2$	28.97	1.40	547	239	6.95	7.04
Ammonia		NH_3	17.03	1.31	1636	731	8.36	9.45
Argon		A	39.94	1.66	705	272	4.97	4.97
Benzene		C_6H_6	78.11	1.12	714	1013	18.43	28.17
Isobutane	iC_4	C_4H_{10}	58.12	1.10	529	735	22.10	31.11
n-Butane	nC_4	C_4H_{10}	58.12	1.09	551	766	22.83	31.09
Isobutylene	iC_4—	C_4H_8	56.10	1.10	580	753	20.44	27.61
Butylene	nC_4—	C_4H_8	56.10	1.11	583	756	20.45	27.64
Carbon dioxide		CO_2	44.01	1.30	1073	548	8.71	10.05
Carbon monoxide		CO	28.01	1.40	510	242	6.96	7.03
Carbureted water gas[b]		—	19.48	1.35	454	235	7.60	8.33
Chlorine		Cl_2	70.91	1.36	1119	751	8.44	8.52
Coke oven gas[b]		—	10.71	1.35	407	197	7.69	8.44
n-Decane	nC_{10}	$C_{10}H_{22}$	142.28	1.03	320	1115	53.67	74.27
Ethane	C_2	C_2H_6	30.07	1.19	708	550	12.13	16.33
Ethyl alcohol		C_2H_5OH	46.07	1.13	927	930	17	21
Ethyl chloride		C_2H_4Cl	64.52	1.19	764	829	14.5	18
Ethylene	C_2—	C_2H_4	28.05	1.24	742	510	10.02	13.41
Flue gas[b]			30.00	1.38	563	264	7.23	7.50
Helium		He	4.00	1.66	33	9	4.97	4.97

(continued)

TABLE 2 (*continued*)

Gas or Vapor	Hydrocarbon Reference Symbols	Chemical Formula	Molecular Mass	Specific Heat Ratio $k = c_p/c_v$ at 60°F	Critical Conditions		$C_{p,m}$ [a]	
					Absolute Pressure p_c (lb/in.²abs)	Absolute Temperature T_c (°R)	At 50°F	At 300°F
n-Heptane	nC_7	C_7H_{16}	100.20	1.05	397	973	39.52	53.31
n-Hexane	nC_6	C_6H_{14}	86.17	1.06	440	915	33.87	45.88
Hydrogen		H_2	2.02	1.41	188	60	6.86	6.98
Hydrogen sulfide		H_2S	34.08	1.32	1306	673	8.09	8.54
Methane	C_1	CH_4	16.04	1.31	673	344	8.38	10.25
Methyl alcohol		CH_3OH	32.04	1.20	1157	924	10.5	14.7
Methyl chloride		CH_3Cl	50.49	1.20	968	750	11.0	12.4
Natural gas[b]		—	18.82	1.27	675	379	8.40	10.02
Nitrogen		N_2	28.02	1.40	492	228	6.96	7.03
n-Nonane	nC_9	C_9H_{20}	128.25	1.04	345	1073	48.44	67.04
Isopentane	iC_5	C_5H_{12}	72.15	1.08	483	830	27.59	38.70
n-Pentane	nC_5	C_5H_{12}	72.15	1.07	489	847	28.27	38.47
Pentylene	$C_5{=}$	C_5H_{10}	70.13	1.08	586	854	25.08	34.46
n-Octane	nC_8	C_8H_{18}	114.22	1.05	362	1025	43.3	59.90
Oxygen		O_2	32.00	1.40	730	278	6.99	7.24
Propane	C_3	C_3H_8	44.09	1.13	617	666	16.82	23.57
Propylene	$C_3{=}$	C_3H_6	42.08	1.15	668	658	14.75	19.91
Blast furnace gas[b]		—	29.6	1.39	—	—	7.18	7.40
Cat cracker gas[b]		—	28.83	1.20	674	515	11.3	15.00
Sulfur dioxide		SO_2	64.06	1.24	1142	775	9.14	9.79
Water vapor		H_2O	18.02	1.33	3208	1166	7.98	8.23

[a]Use straight line interpolation or extrapolation to approximate $C_{p,m}$ (in Btu/mol·°R) at actual inlet T. (For greater accuracy, *average T* should be used.)
[b]Approximate values based on average composition.

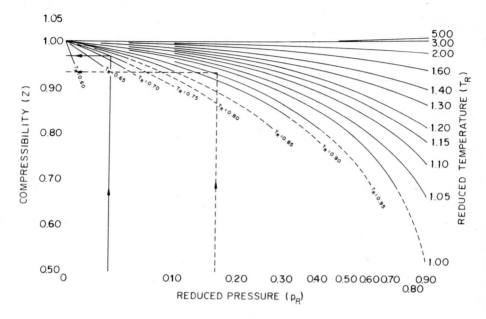

FIG. 1. In English units.

Total hp = gas hp + bearing and oil seal losses

$$\text{Gas hp} = \frac{m \times H_p}{\eta_p \times 33{,}000}$$

Determine losses from Fig. 5, based on type of seal selected.

Step 9.   Find actual discharge temperature (t_2):

$$t_2 = \frac{H_p}{Z_{\text{av}} R \left(\dfrac{k}{k-1} \right) \eta_p} + t_1$$

Step 10.   Calculate discharge flow (Q_2):

TABLE 3 Elliott Compressor Specifications

Frame	Normal Inlet Flow Range[a] (ft³/min)	Nominal Polytropic Head per Stage[b] (H_p)	Nominal Polytropic Efficiency (η_p)	Nominal Maximum No. of Stages[c]	Speed at Nominal Polytropic Head/Stage
29M	500–8,000	10,000	.76	10	11,500
38M	6,000–23,000	10,000/12,000	.77	9	8,100
46M	20,000–35,000	10,000/12,000	.77	9	6,400
60M	30,000–58,000	10,000/12,000	.77	8	5,000
70M	50,000–85,000	10,000/12,000	.78	8	4,100
88M	75,000–130,000	10,000/12,000	.78	8	3,300
103M	110,000–160,000	10,000	.78	7	2,800
110M	140,000–190,000	10,000	.78	7	2,600
25MB (H) (HH)	500–5,000	12,000	.76	12	11,500
32MB (H) (HH)	5,000–10,000	12,000	.78	10	10,200
38MB (H)	8,000–23,000	10,000/12,000	.78	9	8,100
46MB	20,000–35,000	10,000/12,000	.78	9	6,400
60MB	30,000–58,000	10,000/12,000	.78	8	5,000
70MB	50,000–85,000	10,000/12,000	.78	8	4,100
88MB	75,000–130,000	10,000/12,000	.78	8	3,300

[a]Maximum flow capacity is reduced in direct proportion to speed reduction.
[b]Use either 10,000 or 12,000 ft for each impeller where this option is mentioned.
[c]At reduced speed, impellers can be added.

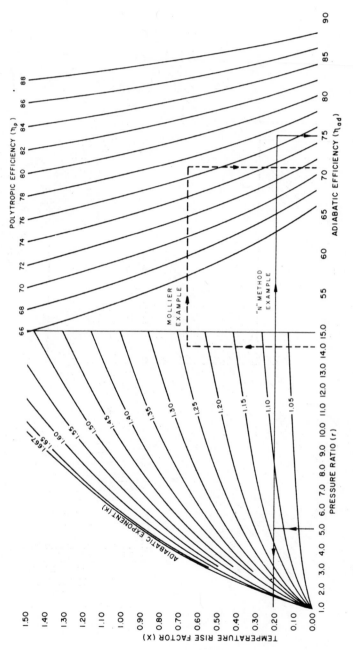

FIG. 2. In English units.

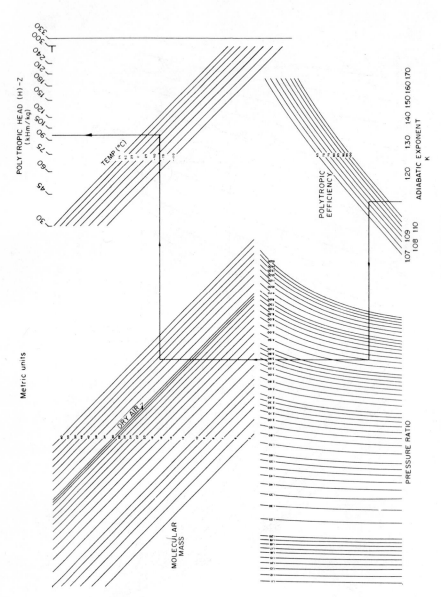

FIG. 3. *N* method generalized selection chart. In metric units.

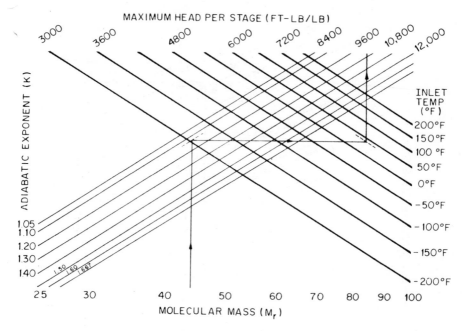

FIG. 4. In English units.

$$Q_2 = Q_1 \frac{p_1}{p_2} \frac{T_2}{T_1} \frac{Z_2}{Z_1}$$

Sample Calculation by the *N* Method (English Units)

Calculate the Elliott compressor required to handle a process gas at the following operating conditions: Inlet temperature $(t_1) = 41°F$, inlet pressure $(p_1) = 20.3$ lb/in.²abs. discharge pressure $(p_2) = 101.5$ lb/in.²abs, gas con-

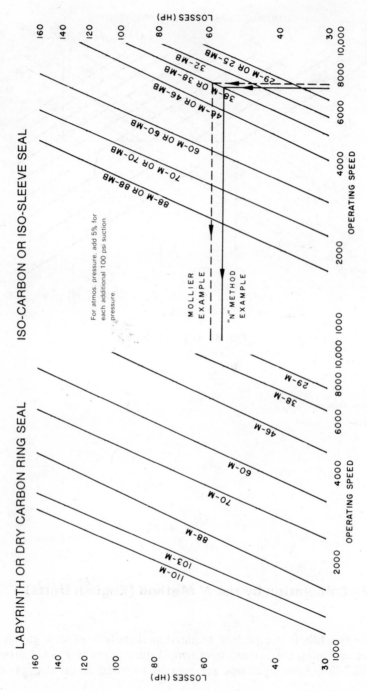

FIG. 5.   In English units.

ditions: 2400 mol/h of mixture of propane (89%), butane (6%), and ethane (5%) (by volume or mol%).

Step 1. Calculate gas mixture properties as shown in Table 4.

Step 2. Calculate inlet flow (Q_1):

$$\text{Mass flow } (m) = \frac{2400 \times 44.23}{60} = 1769 \text{ lb/min}$$

$$p_{R_1} = \frac{20.3}{617.6} = 0.0329$$

$$T_{R_1} = \frac{41 + 460}{666.2} = 0.752$$

From Fig. 1, $Z_1 = 0.97$.

$$v_1 = 0.97 \times \frac{1544}{44.23} \times \frac{(41 + 460)}{144 \times 20.3} = 5.80$$

$$Q_1 = 5.80 \times 1769 = 10,260 \text{ inlet ft}^3/\text{min}$$

Step 3. Select compressor frame. From Table 3, the smallest compressor frame capable of handling this flow (10,260 inlet ft^3/min) is Frame 38M. Note that this is a horizontally-split machine; available up to 9 stages; average polytropic efficiency 77%; 8100 rev/min at 12000 ft head per stage.

Step 4. Calculate average compressibility:

$$r \text{ (pressure ratio)} \frac{p_2}{p_1} = \frac{101.5}{20.3} = 5$$

$$k = 1.135 \text{ (Step 1)}$$

$$\eta_p = 0.77 \text{ (Table 3)}$$

From Fig. 2,

$$X = 0.21$$

$$\eta_{ad} = 0.748$$

$$T_2 \text{ (approx.)} = \frac{0.21(41 + 460)}{0.748} + (41 + 460) = 641.7^\circ R$$

$$T_{R_2} = \frac{T_2}{T_c} = \frac{641.7}{666.2} = 0.963$$

$$p_{R_2} = \frac{p_2}{p_c} = \frac{101.5}{617.6} = 0.164$$

From Fig. 1,

$$Z_2 = 0.93$$

$$Z_{av} = \frac{Z_1 + Z_2}{2} = \frac{0.97 + 0.93}{2} = 0.95$$

Step 5. Calculate polytropic head (H_p). From Fig. 3, knowing

$$k \text{ (Step 1)} = 1.135$$

$$r \ (p_2/p_1) = 5$$

$$t_1 \ (^\circ F) = 41$$

$$\eta_p \text{ (Table 3)} = 0.77$$

Molecular mass (mix) (Step 1) $= 44.23$

$$H_p/Z = 32,000$$

$$Z_{av} = 0.95$$

$$H_p = 30,400 \text{ ft}$$

Or, more accurately, from the equations:

$$\frac{n - 1}{n} = \frac{1.135 - 1}{1.135 \times 0.77} = 0.1545$$

$$H_p = 0.95 \times \frac{1544}{44.23} \times \frac{(41 + 460)}{0.1545} \times (5^{0.1545} - 1) = 30,350 \text{ ft}$$

Step 6. Find number of stages required. From Fig. 4 (knowing molecular mass of mix, k_1, and t_1):

Maximum head per stage $= 10,080 \text{ ft}$

TABLE 4

Gas Mixture	(1) Mol % Each Gas	(2) Mol/h (mol % × 2400)	(3) Molecular Mass (Table 2)	(4) (1) × (3)	(5) Mass % [(4) ÷ 44.23] × 100	(6) T_c, °R (Table 2)	(7) p_c, lb/in.^{2} (Table 2)	(8) (1) × (6)	(9) (1) × (7)	(10) $C_{p,m}$ (Table 2) Btu/mol·°F	(11) (1) × (10)
Propane	89%	2136	44.09	39.24	88.72%	666	617	592.7	549.1	16.58	14.76
n-Butane	6%	144	58.12	3.49	7.89%	766	551	46.0	33.1	22.53	1.35
Ethane	5%	120	30.07	1.50	3.39%	550	708	27.5	35.4	11.98	0.60
		2400		44.23 (Apparent molecular mass of mixture)				666.2 ($T_{c\,(\text{mix})}$)	617.6 ($p_{c\,(\text{mix})}$)		16.71 ($C_{p,m\,(\text{mix})}$)

$$k_{(\text{mixture})} = \frac{16.71}{16.71 - 1.99} = 1.135$$

$$\text{No. of stages} = \frac{30,400}{10,080} = 3.02 \text{ or } 4 \text{ stages}$$

Step 7.   Find speed required:

$$\text{Speed} = 8,100 \sqrt{\frac{30,400}{10,000 \times 4}} = 7,060 \text{ rev/min}$$

(10,000 ft impellers have been selected.)

Step 8.   Find shaft horsepower required:

$$\text{Gas hp} = \frac{1,769 \times 30,400}{0.77 \times 33,000} = 2,116 \text{ hp}$$

Bearing and oil seal loss (from Fig. 5) and assuming isocarbon seals, for Frame 38M compressor = 56 hp.

$$\text{Shaft hp} = 2116 + 56 = 2172 \text{ hp}$$

Step 9.   Find actual discharge temperature (t_2):

$$t_2 = \frac{30,400}{0.95 \times \dfrac{1544}{44.23} \left(\dfrac{1.135}{1.135 - 1} \right) \times 0.77} + 41 = 182.6°F$$

Step 10.   Calculate discharge flow (Q_2):

$$Q_2 = 10,260 \times \frac{20.3}{101.5} \times \frac{182.6 + 460}{41 + 460} \times \frac{0.93}{0.97} = 2,525 \text{ ft}^3/\text{min}$$

Compressor Calculation by the Mollier Method (English Units)

Refer to the simplified Mollier diagram (Fig. 6).

Step 1.   Find inlet flow (Q_1):

$$Q_1 = v_1 \times m$$

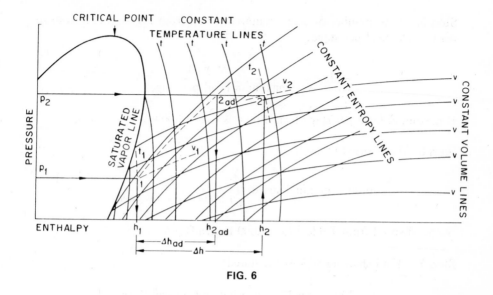

FIG. 6

Locate inlet state point (1) at intersection of p_1 and t_1. Read v_1 by interpolating between specific volume (v) lines.

Step 2. Select compressor frame. Given inlet flow (Q_1), use Table 3.

Step 3. Find adiabatic head (H_{ad}). Read inlet enthalpy (h_1) directly below (1). From (1), follow line of constant entropy to discharge pressure (p_2), locating adiabatic discharge state point (2_{ad}). Read adiabatic enthalpy ($h_{2_{ad}}$) directly below (2_{ad}).

$$\Delta h_{ad} = h_{2_{ad}} - h_1 \ \text{(Btu/lb)}$$

Conversion factor: 778 ft · lb/Btu

$$H_{ad} = \Delta h_{ad} \times 778$$

Step 4. Find polytropic head (H_p):

$$H_p = \frac{H_{ad} \times \eta_p}{\eta_{ad}}$$

Find k from Table 2. Calculate pressure ratio, $r = p_2/p_1$. Find η_p from Table 3 and use Fig. 2 to find η_{ad}.

Step 5.   Find number of stages required. First, from Fig. 4, find *maximum* permissible head per stage:

$$\text{Stages} = \frac{H_p \ (\text{Step 4})}{\text{maximum head per stage}}$$

If maximum head per stage from Fig. 4 is over 12,000 ft, use 12,000.

Step 6.   Find speed required:

$$\text{Speed} = \text{nominal speed} \sqrt{\frac{H_p}{12,000 \times \text{no. of stages}}}$$

Nominal speed from Table 3 (at 12,000 ft. head)

Step 7.   Find shaft horsepower required:

$$\text{Total hp} = \text{gas hp} + \text{bearing and oil seal losses}$$

$$\text{Gas hp} = \frac{m \times H_p}{\eta_p \times 33000}$$

Determine losses from Fig. 5 based on type of seal selected.

Step 8.   Find actual discharge enthalpy (h_2):

$$h_2 = \frac{\Delta h_{ad}}{\eta_{ad}} + h_1$$

H_{ad} from Step 3; η_{ad} from Step 4

Step 9.   Find discharge temperature (t_2) and specific volume (v_2). On Mollier, plot vertically from h_2 (Step 8) to p_2. Read t_2 and v_2.

Step 10.   Find discharge flow (Q_2):

$$Q_2 = m \times v_2$$

Sample Calculation by the Mollier Method (English Units)

Calculate the Elliott compressor required to handle ethylene at the following operating conditions: 90,000 lb/h; inlet temperature $(t_1) = -140°F$; inlet pressure $(p_1) = 15$ lb/in.^{2}abs; discharge pressure $(p_2) = 215$ lb/in.^{2}abs. Refer to the ethylene Mollier diagram (Fig. 7).

Step 1.   Calculate inlet flow (Q_1):

$$\text{Mass flow } (m) = \frac{90,000}{60} = 1,500 \text{ lb/min}$$

From Fig. 7, at p_1 (15 lb/in.^{2}abs) and t_1 $(-140)v_1 = 8.0$:

$$Q_1 = 8.0 \times 1,500 = 12,000 \text{ ft}^3/\text{min}$$

Step 2.   Select compressor frame. From Table 3, the smallest compressor frame capable of handling the flow (12,000 ft^3/min) is Frame 38M. Note that this is a horizontally-split machine available up to 9 stages; average polytropic efficiency 77%; 8100 rev/min at 10,000 ft, or 12,000 ft head per stage.

Step 3.   Find adiabatic head (H_{ad}). From Mollier, inlet enthalpy $(h_1) = 87.5$. Follow line of constant entropy to p_2 (215 lb/in.^{2}abs). Read adiabatic discharge state point $(h_{2_{ad}}) = 169.0$:

$$\Delta h_{ad} = 169 - 87.5 = 81.5$$
$$H_{ad} = 81.5 \times 778 = 63,400 \text{ ft}$$

Step 4.   Find polytropic head (H_p). From Table 2, k for ethylene $= 1.24$. From Table 3,

$$\eta_p = 0.77; \text{ pressure ratio } (r) = \frac{215}{15} = 14.33$$

From Fig. 2, at $r = 14.33$, $k = 1.24$, and $\eta_p = 0.77$; $\eta_{ad} = 0.708$:

$$H_p = \frac{63,400 \times 0.77}{0.708} = 68,950 \text{ ft}$$

Step 5.   Find number of stages required. From Table 2, M_r for ethylene = 28.05. From Fig. 4, using $M_r = 28.05$, $k = 1.24$, $t_1 = -140$; maximum head per stage = 11,200 ft:

$$\text{No. stages} = \frac{68,950}{11,200} = 6.15 \text{ or } 7 \text{ stages}$$

Step 6.   Find speed required:

Nominal speed (Table 3) = 8,100 rev/min

$$8,100 \sqrt{\frac{68,950}{12,000 \times 7}} = 7,340 \text{ rev/min}$$

Step 7.   Find shaft horsepower required:

$$\text{Gas hp} = \frac{1,500 \times 68,950}{0.77 \times 33,000} = 4,070 \text{ hp}$$

Bearing and oil seal losses (from Fig. 5), and assuming isocarbon seals, for Frame 38M compressor = 59 hp.

Shaft hp = 4070 + 59 = 4129 hp

Step 8.   Find actual discharge enthalpy (h_2):

$$h_2 = \frac{81.5}{0.708} + 87.5 = 202.5 \text{ Btu/lb}$$

Step 9.   Find discharge temperature and specific volume. In Fig. 7, plot vertically from h_2 to p_2 (215 lb/in.^{2}abs). Read $t_2 = 195°\text{F}$; $v_2 = 1.14$.

Step 10.   Find discharge flow:

$$Q = 1500 \times 1.14 = 1710 \text{ ft}^3/\text{min}$$

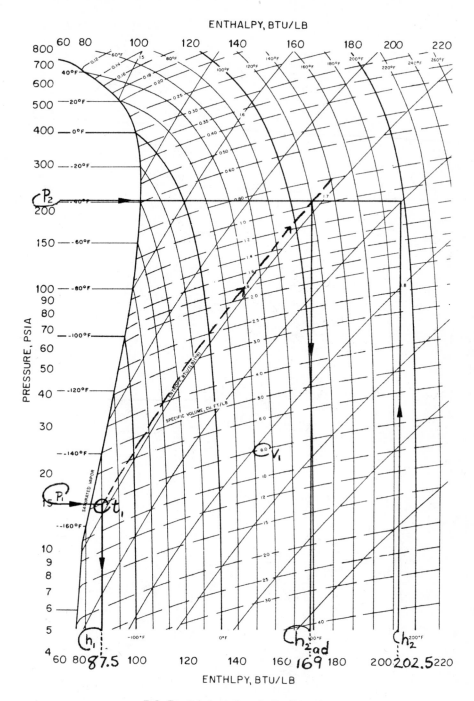

FIG. 7. Ethylene chart. In English units.

TABLE 5

Gas Mixture	(1) Mol% Each Gas	(2) kmol/h Each Gas	(3) Molecular Mass (Table 6)	(4) (1) × (3)	(5) Mass %	(6) T_c (Table 6)	(7) p_c (Table 6)	(8) (1) × (6)	(9) (1) × (7)	(10) $C_{p,m}$ (Table 6)	(11) (1) × (10)
...	...	...	...	a	$a/d \times 100$	...	...	...	...	...	...
...	...	...	...	b	$b/d \times 100$	...	...	...	...	...	...
...	...	...	...	c	$c/d \times 100$	...	...	...	...	...	...
				d (Apparent molecular mass of mixture)				$T_{c\,(mix)}$	$p_{c\,(mix)}$		$C_{p,m\,(mix)}$

Calculate $k_{(mixture)} = \dfrac{C_{p,m\,(mix)}}{C_{p,m\,(mix)} - 8.32}$

Compressor Calculation by the *N* Method (Metric Units)

Step 1. If a *pure* gas, begin with Step 2. If a *mixed* gas, calculate mixture properties as shown in Table 5 (see Table 6).

Step 2. Calculate inlet flow (Q_1):

$$Q_1 = v_1 \times m$$

$$v_1 = \frac{Z_1 R T_1}{p_1 \times 10^5}$$

$$m = \text{kmol/h} \times M_r \text{ (kg/h)}$$

$$R = 8314/M_r$$

Assume Z_1 to be 1 or use Fig. 8. If using Fig. 8:

$$p_{R_1} = p_1/p_c, \qquad T_{R_1} = T_1/T_c$$

Step 3. Select compressor frame. Given inlet volume (Q_1), use Table 3.

Step 4. Calculate average compressibility (Z_{av}):

$$Z_{av} = \frac{Z_1 + Z_2}{2}$$

Z_1 from Step 2. Z_2 as follows:*

$$T_2 \text{ (approx.)} = \frac{X}{\eta_{ad}} (T_1) + T_1$$

From Fig. 9, find X (temperature rise factor) and η_{ad} [using pressure ratio (r) as given, k from Step 1 or Table 6, η_p from Table 7]. Then calculate Z_2 (as in Step 2, using Fig. 8) using p_2 (given) and T_2 as calculated.

Step 5. Calculate polytropic head (H_p).† Using Fig. 3, determine H_p/Z_{av}. Multiply by Z_{av} to obtain head or, for greater accuracy:

---

*This approximate T_2 may differ slightly from the actual discharge temperature since the effect of compressibility upon temperature rise has not yet been considered.
†Adiabatic head may be calculated by using the 100% efficiency line in Fig. 3.

TABLE 6 Gas Properties (most values taken from *Natural Gas Processors Suppliers Association Engineering Data Book—1972*, 9th edition)

Gas or Vapor	Hydrocarbon Reference Symbols	Chemical Formula	Molecular Mass	Specific Heat Ratio $k = c_p/c_v$ at 15.5°C	Critical Conditions Absolute Pressure p_c (bar)	Critical Conditions Absolute Temperature T_c (°K)	$C_{p,m}$[a] At 0°C	$C_{p,m}$[a] At 100°C
Acetylene	$C_2{=}$	C_2H_2	26.04	1.24	62.4	309.4	42.16	48.16
Air		$N_2 + O_2$	28.97	1.40	37.7	132.8	29.05	29.32
Ammonia		NH_3	17.03	1.31	112.8	406.1	34.65	37.93
Argon		A	39.94	1.66	48.6	151.1	20.79	20.79
Benzene		C_6H_6	78.11	1.12	49.2	562.8	74.18	103.52
Isobutane	iC_4	C_4H_{10}	58.12	1.10	36.5	408.3	89.75	116.89
n-Butane	nC_4	C_4H_{10}	58.12	1.09	38.0	425.6	93.03	117.92
Isobutylene	$iC_4{-}$	C_4H_8	56.10	1.10	40.0	418.3	83.36	104.96
Butylene	$nC_4{-}$	C_4H_8	56.10	1.11	40.2	420.0	83.40	105.06
Carbon dioxide		CO_2	44.01	1.30	74.0	304.4	36.04	40.08
Carbon monoxide		CO	28.01	1.40	35.2	134.4	29.10	29.31
Carbureted water gas[b]		—	19.48	1.35	31.3	130.6	31.58	33.78
Chlorine		Cl_2	70.91	1.36	77.2	417.2	35.29	35.53
Coke oven gas[b]		—	10.71	1.35	28.1	109.4	31.95	34.21
n-Decane	nC_{10}	$C_{10}H_{22}$	142.28	1.03	22.1	619.4	218.35	280.41
Ethane	C_2	C_2H_6	30.07	1.19	48.8	305.6	49.49	62.14
Ethyl alcohol		C_2H_5OH	46.07	1.13	63.9	516.7	69.92	81.97
Ethyl chloride		C_2H_4Cl	64.52	1.19	52.7	460.6	59.61	70.16
Ethylene	$C_2{-}$	C_2H_4	28.05	1.24	51.2	283.3	40.90	51.11

Flue gas[b]			30.00	1.38	38.8	146.7	30.17	30.98
Helium		He	4.00	1.66	2.3	5.0	20.79	20.79
n-Heptane	nC$_7$	C$_7$H$_{16}$	100.20	1.05	27.4	540.6	161.20	202.74
n-Hexane	nC$_6$	C$_6$H$_{14}$	86.17	1.06	30.3	508.3	138.09	174.27
Hydrogen		H$_2$	2.02	1.41	13.0	33.3	28.67	29.03
Hydrogen Sulfide		H$_2$S	34.08	1.32	90.0	373.9	33.71	35.07
Methane	C$_1$	CH$_4$	16.04	1.31	46.4	191.1	34.50	40.13
Methyl alcohol		CH$_3$OH	32.04	1.20	79.8	513.3	42.67	55.32
Methyl chloride		CH$_3$Cl	50.49	1.20	66.7	416.7	45.60	49.82
Natural gas[b]		—	18.82	1.27	46.5	210.6	34.66	39.54
Nitrogen		N$_2$	28.02	1.40	33.9	126.7	29.10	29.31
n-Nonane	nC$_9$	C$_9$H$_{20}$	128.25	1.04	23.8	596.1	197.07	253.10
Isopentane	iC$_5$	C$_5$H$_{12}$	72.15	1.08	33.3	461.1	112.09	145.56
n-Pentane	nC$_5$	C$_5$H$_{12}$	72.15	1.07	33.7	470.6	115.21	145.94
Pentylene	C$_5$—	C$_5$H$_{10}$	70.13	1.08	40.4	474.4	102.11	130.37
n-Octane	nC$_8$	C$_8$H$_{18}$	114.22	1.05	25.0	569.4	176.17	226.17
Oxygen		O$_2$	32.00	1.40	50.3	154.4	29.17	29.92
Propane	C$_3$	C$_3$H$_8$	44.09	1.13	42.5	370.0	68.34	88.68
Propylene	C$_3$—	C$_3$H$_6$	42.08	1.15	46.1	365.6	60.16	75.70
Blast furnace gas[b]		—	29.6	1.39	—	—	29.97	30.64
Cat cracker gas[b]		—	28.83	1.20	46.5	286.1	46.16	57.31
Sulfur dioxide		SO$_2$	64.06	1.24	78.7	430.6	38.05	40.00
Water vapor		H$_2$O	18.02	1.33	221.2	647.8	33.31	34.07

[a]Use straight line interpolation or extrapolation to approximate $C_{p,m}$ [in kJ/(kmol·°K)] at actual inlet T. (For greater accuracy, average T should be used.)
[b]Approximate values based on average composition.

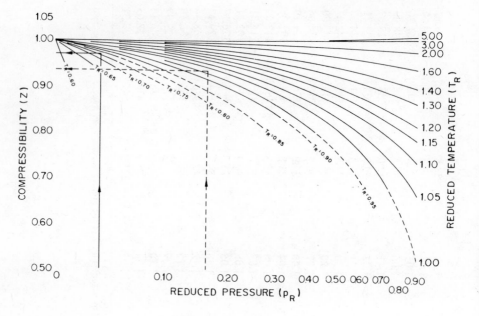

FIG. 8. In metric units.

$$H_p = \frac{Z_{av}RT_1}{1000\dfrac{n-1}{n}}\left[\left(\frac{p_2}{p_1}\right)^{\frac{n-1}{n}} - 1\right]$$

$$\frac{n-1}{n} = \frac{k-1}{k(\eta_p)}$$

η_p from Table 4.

Step 6. Find number of stages required. First, from Fig. 10, find maximum permissible head per stage:

$$\text{Stages} = \frac{H_p \text{ (Step 5)}}{\text{maximum head per stage}}$$

If maximum head per stage from Fig. 10 is over 36 kN · m/kg, use 36.

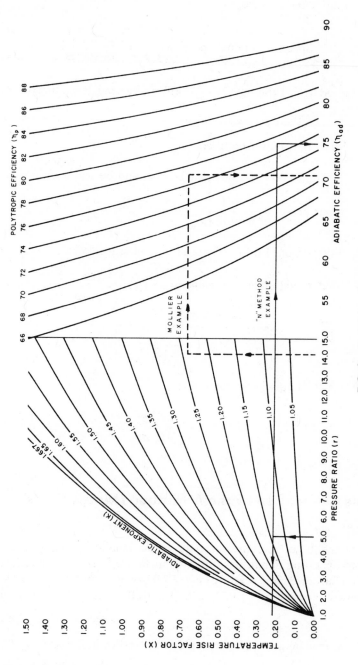

FIG. 9. In metric units.

Step 7.   Find speed required:

$$\text{Speed} = \text{nominal speed} \sqrt{\frac{H_p}{36 \times \text{no. of stages}}}$$

Nominal speed from Table 4 (at 30 or 36 kN $\cdot$ m/kg head depending on impeller selected).

Step 8.   Find shaft power required:

$$\text{Total kW} = \text{gas power} + \text{bearing and oil seal losses}$$

$$\text{Gas power} = \frac{mH_p}{3600 \times \eta_p}$$

Determine losses from Fig. 11 based on type of seal selected.

Step 9.   Find actual discharge temperature (t_2):

$$t_2 = \frac{1000 H_p}{Z_{\text{av}} R \left(\dfrac{k}{k-1} \right) \eta_p} + t_1$$

Step 10. Calculate discharge flow (Q_2):

$$Q_2 = Q_1 \frac{p_1}{p_2} \frac{T_2}{T_1} \frac{Z_2}{Z_1}$$

Sample Calculation by the *N* Method (Metric Units)

Calculate the Elliott compressor required to handle a process gas at the following operating conditions: Inlet temperature (t_1) = 5°C; inlet pressure (p_1) = 1.4 bar; discharge pressure (p_2) = 7.0 bar; gas conditions: 1090 kmol/h of mixture of propane (89%), butane (6%), and ethane (5%) (by volume or mol-%).

Step 1.   Calculate gas mixture properties as shown in Table 8.

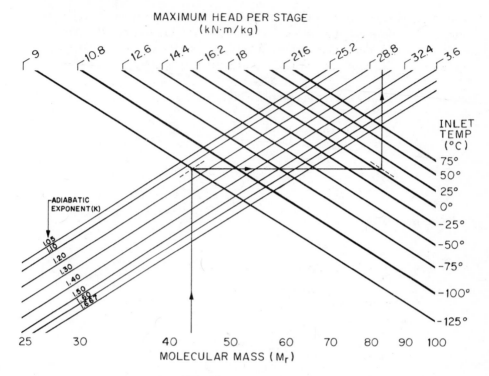

MAXIMUM HEAD PER STAGE
(kN·m/kg)

FIG. 10. In metric units.

Step 2. Calculate inlet flow (Q_1):

$$\text{Mass flow } (m) = 1{,}090 \times 44.23 = 48{,}211 \text{ kg/h}$$

$$P_{R_1} = \frac{1.4}{42.5} = 0.0329$$

$$T_{R_1} = \frac{5 + 273.1}{370.1} = 0.752$$

From Fig. 8, $Z_1 = 0.97$.

$$v_1 = 0.97 \times \frac{8{,}314}{44.23} \times \frac{(5 + 273)}{1.4 \times 10^5} = 0.362 \text{ m}^3/\text{kg}$$

$$Q_1 = 0.362 \times 48{,}211 = 17{,}450 \text{ m}^3/\text{h}$$

Step 3.   Select compressor frame. From Table 7, the smallest compressor frame capable of handling this flow (17,450 inlet m^3/h) is Frame 38M. Note that this is a horizontally-split machine; available up to 9 stages; average polytropic efficiency 77%; 8100 rev/min at 36 kN · m/kg head per stage.

Step 4.   Calculate average compressibility:

$$r \text{ (pressure ratio)} \frac{p_2}{p_1} = \frac{7.0}{1.4} = 5$$

$$k = 1.135 \text{ (Step 1)}$$

$$\eta_p = 0.77 \text{ (Table 7)}$$

From Fig. 9,

$$X = 0.21$$

$$\eta_{ad} = 0.748$$

$$T_2 \text{ (approx.)} = \frac{0.21(5 + 273.1)}{0.748} + (5 + 273.1) = 356.3 \text{ K}$$

$$T_{R_2} = \frac{T_2}{T_c} = \frac{356.3}{370.1} = 0.963$$

$$p_{R_2} = \frac{p_2}{p_c} = \frac{7.0}{42.5} = 0.164$$

From Fig. 8,

$$Z_2 = 0.93$$

$$Z_{av} = \frac{Z_1 + Z_2}{2} = \frac{0.97 + 0.93}{2} = 0.95$$

Step 5.   Calculate polytropic head (H_p). From Fig. 3, knowing

$$k \text{ (Step 1)} = 1.135$$

$$r \ (p_2/p_1) = 5$$

$$t_1 \ (^\circ C) = 5$$

$$\eta_p \text{ (Table 7)} = 0.77$$

$$\text{Molecular mass (mix) (Step 1)} = 44.23$$

TABLE 7 Elliott Compressor Specifications

Frame	Normal Inlet Flow Range[a] (m³/h)	Nominal Polytropic Head per Stage[b] (H_p)	Nominal Polytropic Efficiency (η_p)	Nominal Maximum No. of Stages[c]	Speed at Nominal Polytropic Head/Stage
29M	850–13,600	30	.76	10	11,500
38M	10,000–39,000	30/36	.77	9	8,100
46M	34,000–60,000	30/36	.77	9	6,400
60M	51,000–99,000	30/36	.77	8	5,000
70M	85,000–145,000	30/36	.78	8	4,100
88M	125,000–220,000	30/36	.78	8	3,300
103M	185,000–270,000	30	.78	7	2,800
110M	235,000–320,000	30	.78	7	2,600
25MB (H) (HH)	850–8,500	36	.76	12	11,500
32MB (H) (HH)	8,500–17,000	36	.78	10	10,200
38MB (H)	13,600–39,000	30/36	.78	9	8,100
46MB	34,000–60,000	30/36	.78	9	6,400
60MB	51,000–99,000	30/36	.78	8	5,000
70MB	85,000–145,000	30/36	.78	8	4,100
88MB	125,000–220,000	30/36	.78	8	3,300

[a]Maximum flow capacity is reduced in direct proportion to speed reduction.
[b]Use either 30 or 36 kN · m/kg for each impeller where this option is mentioned.
[c]At reduced speed, impellers can be added.

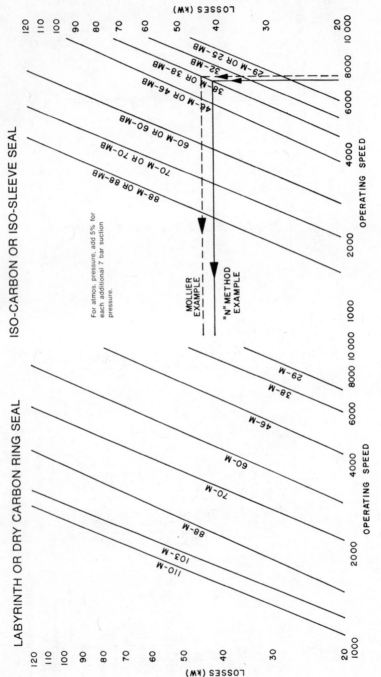

FIG. 11. In metric units.

TABLE 8

Gas Mixture	(1) Mol % Each Gas	(2) kmol/h (mol% × 1090)	(3) Molecular Mass (Table 6)	(4) (1) × (3)	(5) Mass % [(4) ÷ 44.23] × 100	(6) T_c, °K (Table 6)	(7) p_c, bar (Table 6)	(8) (1) × (6)	(9) (1) × (7)	(10) $C_{p,m}$ (Table 6) kJ/(kmol · °K)	(11) (1) × (10)
Propane	89%	970.1	44.09	39.24	88.72%	370.0	42.5	329.3	37.8	69.36	61.73
n-Butane	6%	65.4	58.12	3.49	7.89%	425.6	38.0	25.5	2.3	94.27	5.66
Ethane	5%	54.5	30.07	1.50	3.39%	305.6	48.8	15.3	2.4	50.12	2.51
		1090.0		44.23				370.1	42.5		69.90
				(Apparent molecular mass of mixture)				($T_{c\,(\text{mix})}$)	($p_{c\,(\text{mix})}$)		($C_{p,m\,(\text{mix})}$)

$$k_{(\text{mixture})} = \frac{69.90}{69.90 - 8.32} = 1.135$$

$$H_p/Z = 95.65 \text{ kN} \cdot \text{m/kg}$$
$$Z_{av} = 0.95$$
$$H_p = 90.87 \text{ kN} \cdot \text{m/kg}$$

Or, more accurately, from the equations:

$$\frac{n-1}{n} = \frac{1.135 - 1}{1.135 \times 0.77} = 0.1545$$

$$H_p = \frac{0.95}{1000} \times \frac{8314}{44.23} \times \frac{(5 + 273.1)}{0.1545} \times (5^{0.1545} - 1) = 90.72 \text{ kN} \cdot \text{m/kg}$$

Step 6. Find number of stages required. From Fig. 10 (knowing molecular mass of mix, k_1 and t_1):

$$\text{Maximum head per stage} = 30.1 \text{ kN} \cdot \text{m/kg}$$

$$\text{No. of stages} = \frac{90.87}{30.1} = 3.02 \text{ or 4 stages}$$

Step 7. Find speed required:

$$\text{Speed} = 8100 \sqrt{\frac{90.87}{30 \times 4}} = 7060 \text{ rev/min}$$

(30 kN · m/kg impellers have been selected.)

Step 8. Find shaft power required:

$$\text{Gas power} = \frac{48,211 \text{ kg/h}}{3,600 \text{ s/h}} \times \frac{90.87 \text{ kJ/kg}}{0.77} = 1580 \text{ kW}$$

Bearing and oil seal loss (from Fig. 11) and assuming isocarbon seals, for Frame 38M compressor = 42 kW.

$$\text{Shaft power} = 1580 + 42 = 1622 \text{ kW}$$

Step 9. Find actual discharge temperature (t_2):

$$t_2 = \frac{90{,}870}{0.95 \times \dfrac{8{,}314}{44.23} \left(\dfrac{1.135}{1.135 - 1}\right) \times 0.77} + 5 = 83.6°C$$

Step 10.   Calculate discharge flow (Q_2):

$$Q_2 = 17{,}450 \times \frac{1.4}{7} \times \frac{83.6 + 273}{5 + 273} \times \frac{0.93}{0.97} = 4{,}292 \ \text{m}^3/\text{h}$$

Compressor Calculation by the Mollier Method (Metric Units)

Refer to the simplified Mollier diagram (Fig. 6).

Step 1.   Find inlet flow (Q_1):

$$Q_1 = v_1 \times m$$

Locate inlet state point (1) at intersection of p_1 and t_1. Read v_1 by interpolating between specific volume (v) lines.

Step 2.   Select compressor frame. Given inlet flow (Q_1), use Table 7.

Step 3.   Find adiabatic head (H_{ad}). Read inlet enthalpy (h_1) directly below (1). From (1), follow line of constant entropy to discharge pressure (p_2), locating adiabatic discharge state point (2_{ad}). Read adiabatic enthalpy ($h_{2_{ad}}$) directly below (2_{ad}).

$$\Delta h_{ad} = h_{2_{ad}} - h_1 \ (\text{kJ/kg})$$

$$H_{ad} = \Delta h_{ad} \ (\text{kN} \cdot \text{m/kg})$$

Step 4.   Find polytropic head (H_p):

$$H_p = \frac{H_{ad} \times \eta_p}{\eta_{ad}}$$

Find k from Table 6. Calculate pressure ratio, $r = p_2/p_1$. Find η_p from Table 7 and use Fig. 9 to find η_{ad}.

Step 5. Find number of stages required. First, from Fig. 10 find *maximum* permissible head per stage:

$$\text{Stages} = \frac{H_p \text{ (Step 4)}}{\text{maximum head per stage}}$$

If maximum head per stage from Fig. 10 is over 36 kN · m/kg, use 36.

Step 6. Find speed required:

$$\text{Speed} = \text{nominal speed} \sqrt{\frac{H_p}{36 \times \text{no. of stages}}}$$

Nominal speed from Table 7 (at 36 kN · m/kg head).

Step 7. Find shaft power required:

$$\text{Total kW} = \text{gas power} + \text{bearing and oil seal losses}$$
$$\text{Gas power} = \frac{m}{3600} \times \frac{H_p}{\eta_p}$$

Determine losses from Fig. 11 based on type of seal selected.

Step 8. Find actual discharge enthalpy (h_2):

$$h_2 = \frac{\Delta h_{ad}}{\eta_{ad}} + h_1$$

H_{ad} from Step 3; η_{ad} from Step 4.

Step 9. Find discharge temperature (t_2) and specific volume (v_2). On Mollier, plot vertically from h_2 (Step 8) to p_2. Read t_2 and v_2.

Step 10. Find discharge flow (Q_2):

$$Q_2 = m \times v_2$$

Sample Calculation by the Mollier Method (Metric Units)

Calculate the Elliott compressor required to handle ethylene at the following operating conditions: 40,800 kg/h; inlet temperature $(t_1) = -95.5°C$; inlet pressure $(p_1) = 1.035$ bar; discharge pressure $(p_2) = 14.83$ bars. Refer to the ethylene Mollier diagram (Fig. 12).

Step 1.   Calculate inlet flow (Q_1):

$$\text{Mass flow } (m) = 40,800 \text{ kg/h}$$

From Fig. 12, at p_1 (1.035 bars) and t_1 (-95.5), $v_1 = 0.499$:

$$Q_1 = 0.499 \times 40,800 = 20,375 \text{ m}^3/\text{h}$$

Step 2.   Select compressor frame. From Table 7, the smallest compressor frame capable of handling the flow (20,375 m³/h) is Frame 38M. Note that this is a horizontally-split machine available up to 9 stages; average polytropic efficiency 77%; 8100 rev/min at 30 kN · m/kg head per stage.

Step 3.   Find adiabatic head (H_{ad}). From Mollier, inlet enthalpy $(h_1) = 203.4$. Follow line of constant entropy to p_2 (14.83 bars). Read adiabatic discharge state point $(h_{2_{ad}}) = 392.8$:

$$\Delta h_{ad} = 392.8 - 203.4 = 189.4$$
$$H_{ad} = 189.4 \text{ kN} \cdot \text{m/kg}$$

Step 4.   Find polytropic head (H_p). From Table 6, k for ethylene = 1.24. From Table 7,

$$\eta_p = 0.77; \text{ pressure ratio } (r) = \frac{14.83}{1.035} = 14.33$$

From Fig. 9, at $r = 14.33$, $k = 1.24$, and $\eta_p = 0.77$; $\eta_{ad} = 0.708$:

ENTHALPY, kJ/kg

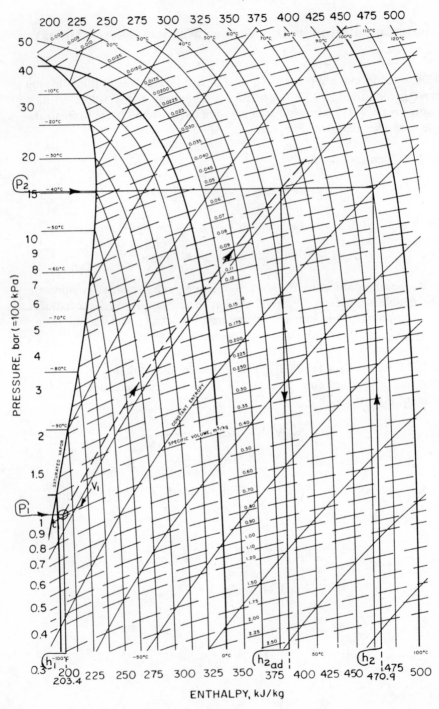

FIG. 12. Ethylene chart. In metric units.

$$H_p = \frac{189.4 \times 0.77}{0.708} = 206 \text{ kN} \cdot \text{m/kg}$$

Step 5. Find number of stages required. From Table 6, M_r for ethylene $= 28.05$. From Fig. 10, using $M_r = 28.05$, $k = 1.24$, $t_1 = -95.5$; maximum head per stage $= 33.5$ kN $\cdot$ m/kg.

$$\text{No. stages} = \frac{206}{33.5} = 6.15 \text{ or } 7 \text{ stages}$$

Step 6. Find speed required:

$$\text{Nominal speed (Table 7)} = 8100 \text{ rev/min}$$

$$8100 \sqrt{\frac{206}{36 \times 7}} = 7340 \text{ rev/min}$$

Step 7. Find shaft power required:

$$\text{Gas power} = \frac{40,800}{3,600} \times \frac{206}{0.77} = 3,030 \text{ kW}$$

Bearing and oil seal losses (from Fig. 11) and assuming isocarbon seals, for Frame 38M compressor $= 44$ kW.

$$\text{Shaft power} = 3030 + 44 = 3074 \text{ kW}$$

Step 8. Find actual discharge enthalpy (h_2):

$$h_2 = \frac{189.4}{0.708} + 203.4 = 470.9 \text{ kJ/kg}$$

Step 9. Find discharge temperature and specific volume. In Fig. 12, plot vertically from h_2 to p_2 (14.83 bars). Read $t_2 = 90.5°C$; $v_2 = 0.0712 \text{ m}^3/\text{kg}$.

Step 10. Find discharge flow:

$$Q = 40,800 \times 0.0712 = 2,905 \text{ m}^3/\text{h}$$

Symbols

$C_{p,m}$	molal specific heat at constant pressure
H	head (ft $\cdot$ lb$_f$/lb$_m$; kN $\cdot$ m/kg)
h	enthalpy (Btu/lb; kJ/kg)
k	adiabatic exponent (c_p/c_v)
m	mass flow (lb/min; kg/h)
M_r	(relative) molecular mass
n	polytropic exponent
p	pressure (lb/in.^{2}abs; bar)
Q	capacity (ft^3/min; m^3/h)
R	gas constant [(1544/M_r)(ft $\cdot$ lb/lb $\cdot$ °F); (8314/M_r)J/(kg $\cdot$ °K)]
r	pressure ratio (p_2/p_1)
T	absolute temperature (°Rankine = °F + 459.6; °Kelvin = °C + 273.15)
t	temperature (°F; °C)
v	specific volume (ft^3/lb; m^3/kg)
X	temperature rise factor
Z	compressibility factor
η	efficiency

Subscripts

ad	adiabatic process
c	ritical
p	polytropic process
R	reduced property
1	inlet conditions
2	discharge conditions

Part B. Reciprocating Compressors

Calculations

General

Solving compressor problems by the use of the ideal gas laws has been reduced to a relatively simple sequence of applying a few basic formulas and obtaining values from curves. In order to help in the understanding of these terms and equations, the first portion of this section covers some derivations.

Basic Reciprocating Compressor Cycle

A graphical representation of the compressor cycle is shown on a pressure–time diagram (Fig. 13).

At the first position the piston is just starting the compression stroke, moving to the left. There is a volume of gas V_1 trapped in the cylinder at a pressure P_1. The gas is compressed approximately adiabatically according to the relation $PVk = \beta$, where k is the isentropic exponent which normally remains nearly constant throughout the compression process.

At the second position the gas has been compressed to a volume V_2 and the pressure has reached P_2 which is equal to the discharge pressure. Thus the discharge valve opens at discharge pressure and, during the remainder of the compression stroke, gas is forced into the discharge line with the pressure remaining constant, as indicated by the path from 2 to 3 on the diagram.

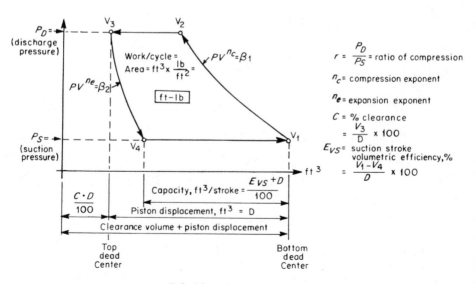

FIG. 13. Ideal PV diagram.

At Position 3 the piston has reached the end of the compression stroke and is just starting the expansion stroke. Thus the pressure drops below the discharge line pressure and the discharge valve closes. Then, as the piston moves to the right, the remaining trapped gas expands along path 3–4, which again is approximately an adiabatic process given by $PVk = \beta$.

At Position 4 the gas has expanded sufficiently so that the pressure has dropped to the suction line pressure. Thus, when the pressure in the cylinder drops below the suction pressure, the inlet valve is forced open and gas enters the cylinder. This continues throughout the remainder of the expansion stroke until Position 1 is reached again. At this point the piston has reached the end of its suction stroke and has reversed direction. As compression starts, the cylinder pressure exceeds suction pressure and the suction valve closes and the process is repeated.

Thus this is the basic operation of a reciprocating compressor with a certain mass of gas being compressed from the suction pressure to the discharge pressure. For a particular design problem the compressor or compressor dimensions and operating speed will have to be sized to deliver the desired flow rate of gas.

Derivation of Suction Volumetric Efficiency

$$V_1 = D + \frac{C}{100} D$$

$$P_D V_2{}^{n_c} = P_S V_1{}^{n_c}$$

$$V_2{}^{n_c} = \frac{P_S}{P_D} V_1{}^{n_c}$$

$$V_2 = V_1 \left[\frac{P_S}{P_D} \right]^{1/n_c}$$

$$V_3 = \frac{C}{100} D$$

$$V_4 = \left(D + \frac{C}{100} D \right) - \frac{E_{VS}}{100} D$$

$$E_{VS} D = 100 V_1 - 100 V_4$$

but

$$V_1 = D + \frac{C}{100} D$$

and

$$V_4 = V_3 \left[\frac{P_D}{P_S} \right]^{1/n_e} = \frac{C}{100} D\,[r]^{1/n_e}$$

therefore

$$E_{VS}D = 100D + CD - CD[r]^{1/n_e}$$

$$E_{VS} = 100 + C - C[r]^{1/n_e}$$

For the ideal case the with $n_e = k$ and collecting terms:

$$E_{VS} = 100 - \frac{P_2^{1/k}}{P_1} - 1 \tag{1}$$

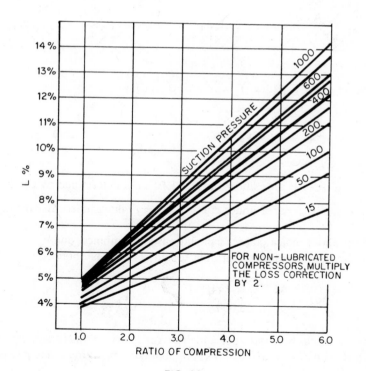

FOR NON–LUBRICATED COMPRESSORS, MULTIPLY THE LOSS CORRECTION BY 2.

FIG. 14

In actual machines, other factors affect the volumetric efficiency. These are leakage across the valves, across the piston rings, and through the packing, plus the heating effect on the incoming gas from the residual heat in the cylinder. In addition, the compressibility of the gas may be included. A better formula has been developed:

$$E_{VS} = 100 - L - C\frac{Z_1}{Z_2}\frac{P_2^{1/k}}{P_1} - 1 \tag{2}$$

where L is taken from Fig. 14.

Work per Cycle = Area of *PV* Diagram   (Refer to Fig. 13)

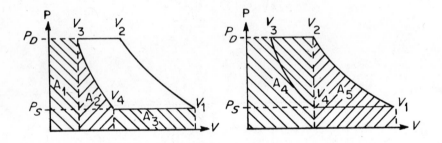

Work/cycle $= (A_4 + A_5) - (A_1 + A_2 + A_3)$

$A_1 = P_D \times V_3$

$A_4 = P_D \times V_2$

$A_3 = P_S \times (V_1 - V_4)$

$A_2 =$ area under curve $PV^{n_e} = \beta_2$, determined by calculus

$A_5 =$ area under curve $PV^{n_e} = \beta$, determined by calculus

When the above indicated area calculations are combined, and $n_c = n_e = k$ for the ideal case, the standard formula for horsepower is obtained.

$$hp_{ad} = \frac{ACFM \times 144 \times P_1}{33,000} \times \frac{k}{k-1}\left[\left(\frac{P_2}{P_1}\right)^{(k-1)/k} - 1\right] \tag{3}$$

This equation has been modified and drawn as a curve, using capacity in million cubic feet per day at 14.6 lb/in.^{2}abs and suction temperature. This curve is given in Fig. 16.

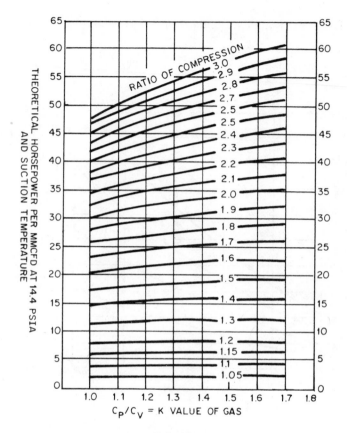

FIG. 16

This horsepower is theoretical and must be modified by compressibility, valve efficiency, and mechanical efficiency.

Valve efficiency allows for the pressure drop across the valves, which gives higher discharge and lower suction pressure within the cylinder than is supplied to the compressor. This may be found from Figs. 17 and 18.

Mechanical efficiency is based on the power loss in the crankcase, and efficiency is usually accepted as 95%. Then, actual horsepower may be calculated as follows:

$$\text{hp actual} = \frac{\text{MM SCFD}}{Z_{std}} \times \frac{T_1}{T_{std}} \times \frac{\text{Bhp}}{\text{MM CFD}} \times \frac{Z_1 + Z_2}{2} \dots$$

$$\times \frac{1}{\text{valve efficiency}} \times \frac{1}{\text{mechanical efficiency}} \quad (4)$$

where MM SCFD is million standard cubic feet per day at 14.4 lb/in.^{2}abs and 60°F.

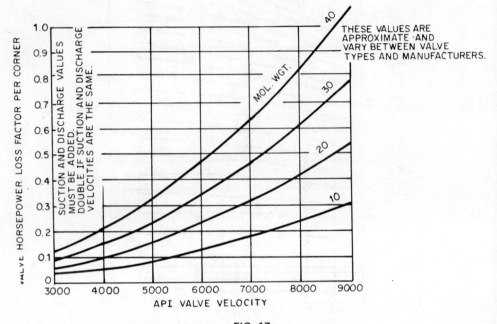

FIG. 17

For approximate calculations an estimated compressor efficiency from Fig. 19 can be used, where E_{ad} = adiabatic efficiency including mechanical losses:

$$\text{hp} = \frac{\text{ACFM} \times 144 \times P_1}{33,000 \times \dfrac{k-1}{k}} \left[\left(\frac{P_2}{P_1}\right)^{(k-1)/k} - 1 \right] \frac{1}{\varepsilon_{ad}} \qquad (5)$$

This equation is used for finding horsepower before sizing the machine. For existing machines, displacement is known and volumetric efficiency can be calculated.

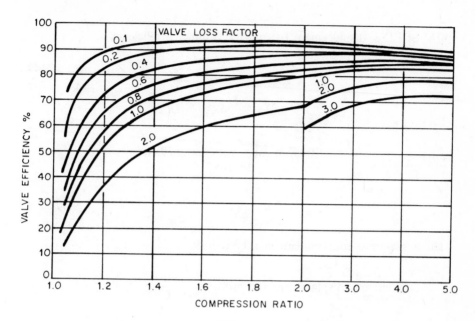

FIG. 18

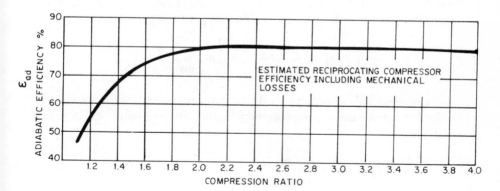

FIG. 19

$$\text{MM SCFD} = \text{displacement (ft}^3/\text{min)} \times \text{volumetric efficiency} \times \frac{T_{std}}{T_1} \cdots$$

$$\times \frac{Z_{std}}{Z_1} \times \frac{P_1}{14.4} \times \frac{1440}{10^6}$$

Then,

$$\text{MM SCFD} \times \frac{T_1}{T_{std}} = \frac{\text{displacement} \times \text{volumetric efficiency}}{10^4} \cdots$$

$$\times \frac{Z_{std}}{Z_1} \times P_1$$

The horsepower equation can then be written

$$\text{hp actual} = \text{displacement (ft}^3/\text{min)} \times \frac{\text{Bhp}}{\text{MM CFD}} \times \frac{E_{VS} \times P_1}{10^4} \cdots$$

$$\times \frac{Z_1 + Z_2}{2Z_1} \times \frac{1}{\text{valve efficiency}} \times \frac{1}{\text{mechanical efficiency}} \quad (6)$$

Machines may be sized on the basis shown in Table 9. Different manufacturers may vary slightly from these sizes.

Reciprocating compressor calculations are done by stage. Therefore, the number of stages must be determined first. This is based on limiting temperature and rod load. In addition, horsepower savings can be made by having the same ratio for each stage. This ratio can be calculated by

$$\text{Ratio per stage} = (\text{overall ratio}) \frac{1}{\text{no. of stages}}$$

Generally speaking, in process work the ratio per stage seldom exceeds 3. Except for extreme cases, compressibility factor has a very minor effect and is usually ignored.

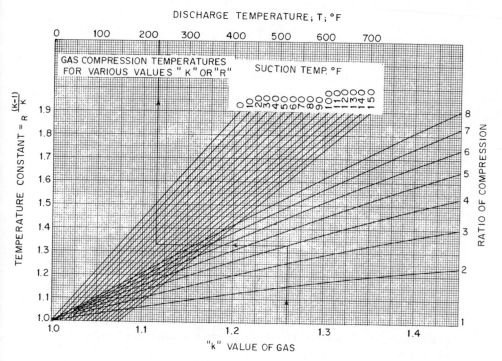

FIG. 20. This chart permits determination of the theoretical or adiabatic compression temperatures of gases from the formula $T_2 = T_1 R^{(k-1)/k}$, where T_2 and T_1 are expressed as absolute temperatures, degrees Rankine. To find T_1, °F, subtract 460 from T_2. The k value of the gas can be obtained from the known, calculated, or estimated specific gravity or percentage molecular weight of the fractions constituting the mixture of gases under consideration. (See Example.) After locating the k value on the abscissa of the above chart, it can be followed vertically to the compression ratio, then left horizontally to the suction temperature, then vertically to the discharge temperature. Arrows show direction through chart. Chart by G. R. Davis.

TABLE 9

Strokes	Horsepower per Stroke	Maximum hp per Machine	Rod Size	Rod Load	Speed
9	250	500	$2\frac{1}{4}$	20,000	600
12	300	1,000	$2\frac{1}{2}$	30,000	400
14	500	1,500	3	45,000	327
16	750	3,500	$3\frac{3}{4}$	75,000	300
18	1,000	6,000	4	100,000	277

Additional Formulas and Graphs

Discharge Temperature

The adiabatic temperature at discharge is given by

$$T_2 = \frac{P_2^{(k-1)/k}}{P_1} T_1 \tag{7}$$

Figure 20 is a graphical solution of Eq. (7).

Cooling jackets, lubricant, and packing and ring material limit cylinder temperatures. 275°F is set as the ideal maximum, with 375°F as the absolute limit of operating temperatures.

Effect of Water Vapor

If a compressor must handle a specified volume of dry gas per minute, the volume of gas plus water vapor can be determined by multiplying the dry gas volume/minute by the factor

$$M = 1 + \frac{P}{P - P_V}$$

where P = pressure of mixture
P_V = pressure of water vapor

P_V can be found from saturated steam table data multiplied by relative humidity.

TABLE 10

Constituent	Mol% Dec.	Molecular Weight		Mol% Dec.	k	
Air	0.276 ×	29	= 8.01	0.276 × 1.400	= 0.386
Methane (CH_4)	0.215 ×	16	= 3.44	0.215 × 1.296	= 0.279
Ethane (C_2H_6)	0.012 ×	30	= 0.36	0.012 × 1.206	= 0.014
Propane (C_3H_8)	0.050 ×	44	= 2.20	0.050 × 1.152	= 0.058
Isobutane (C_4H_{10})	0.071 ×	58	= 4.12	0.071 × 1.110	= 0.079
n-Butane (C_4H_{10})	0.193 ×	58	= 11.40	0.193 × 1.110	= 0.214
Isopentane (C_5H_{12})	0.083 ×	72	= 5.97	0.083 × 1.090	= 0.091
n-Pentane (C_5H_{12})	0.038 ×	72	= 5.18	0.038 × 1.090	= 0.041
Cyclopentane (C_7H_{16})	0.062 ×	100	= 6.20	0.062 × 1.075	= 0.067

Average molecular weight = 46.88   Average k = 1.229

Example Problem

See Table 10 for the average molecular weight and k value for a gas mixture.

Piston Displacement

Piston displacement is the actual volume displaced by the piston as it travels the length of its stroke from Position 1, bottom dead center, to Position 3, top dead center. Piston displacement is normally expressed as the volume displaced per minute or cubic feet per minute. In the case of the double-acting cylinder, the displacement of the crank end of the cylinder is also included. The crank end displacement is, of course, less than the head end displacement by the amount that the piston rod displaced. For a single-acting cylinder:

$$D = \frac{A_{HE} \times S \times \text{rev/min}}{1728}$$

where A_{HE} = area head end of piston, in.^{2}
   S = stroke, in.
   D = piston displacement, ft^3/min

For double-acting cylinders:

$$D_{DA} = \frac{A_{HE} \times S \times \text{rev/min}}{1728} + \frac{A_{CE} \times S \times \text{rev/min}}{1728}$$

or

$$= \frac{(S)(\text{rev/min})(2)}{1728}\left(A_{HE} - \frac{1}{2}A_R\right)$$

where A_R = area rod, in.^{2}

Rod Load

The design compression and tension load on the piston rod must not be exceeded, and therefore the rod load must be checked on each cylinder application. Rod load is defined as

$$R.L. = P_2 \times A_{HE} - P_1 \times A_{CE} \tag{8}$$

where $R.L.$ = rod load in compression in pounds
$\quad A_{HE}$ = cylinder area at head end in any one cylinder
$\quad A_{CE}$ = cylinder area at crank end, usually $A_{CE} = A_{HE}$ − area rod
$\quad P_2$ = discharge pressure, lb/in.^{2}abs
$\quad P_1$ = suction pressure, lb/in.^{2}abs

The limiting rod loads are set by the manufacturer. Some manufacturers require lower values on the rod in tension than in compression. For this, the above limits are checked by reversing A_{HE} and A_{CE}.

Sample Calculation

New Machine.   Consider a new compressor application:

Gas molecular weight	17.76
k value	1.26
Suction temperature	60°F
Suction pressure	150 lb/in.^{2}abs
Discharge pressure	1000 lb/in.^{2}abs
Flow rate	9 MM SCFD
Compression ratio	6.67

From Fig. 20, the discharge temperature is 308°F. The specific gravity of gas = 17.76/28.97 = 0.612. From tables:

$$Z_S = 0.98$$

$$Z_D = 0.97$$

$$Z_{av} = 0.975$$

$$\frac{k-1}{k} = \frac{0.26}{1.26} = 0.206$$

$$\text{Capacity} = 9 \text{ MM SCFD} \times \frac{10^6}{1440} \times \frac{P_{std}}{P_S} \times \frac{T_S}{T_{std}} \times \frac{1}{Z_S}$$

$$= 9 \times \frac{10^6}{1440} \times \frac{14.7}{150} \times \frac{520}{520} \times \frac{1}{0.98}$$

$$= 612 \text{ ACFM}$$

Compression efficiency ε_{ad} from Fig. 19 = 80%.
From Eq (5):

$$\text{hp} = \frac{\text{ACFM} \times 144 \times P_1}{33,000 \times \dfrac{k-1}{k}} \left[\left(\frac{P_2}{P_1} \right)^{(k-1)/k} - 1 \right] \frac{1}{\varepsilon_{ad}}$$

$$= \frac{612 \times 144 \times 150}{33,000 \times 0.206} [6.67^{0.206} - 1] \frac{1}{0.80}$$

$$= 1160$$

1160 hp is too large for one throw from the table, and a ratio of 6.67 is larger than the normal 3, so this machine is best built two stage. This would be a two-cylinder, two-stage machine, 14 in. stroke, 327 rev/min, 45,000 lb rod load.

$$\text{Ratio per stage} = (6.67)^{1/2}$$

$$= 2.58$$

$$\text{Interstage pressure} = 150 \times 2.58 = 386 \text{ lb/in.}^2\text{abs}$$

Interstage temperature from Fig. 9 = 172°F

Capacity of second stage:

$$9 \times \frac{10^6}{1440} \times \frac{14.4}{386} \times \frac{632}{520} \times \frac{1}{0.98} = 288 \text{ ACFM}$$

Assume a clearance of 15%.
Volumetric efficiency:

$$E_{VS} = 100 - L - \% \text{ clearance}\left[\left(\frac{P_2}{P_1}\right)^{1/k} - 1\right]$$

L can be found from Fig. 14.

First stage $E_{VS} = 100 - 6.3 - 15[(2.58)^{1/1.26} - 1] = 76.9\%$

Second stage $E_{VS} = 100 - 7.0 - 15[(2.58)^{1/1.26} - 1] = 76.2\%$

$$\text{Displacement} = 2 \times \frac{\text{area of cylinder}}{144} \times \frac{\text{stroke}}{12} \times \text{rev/min} \times E_{VS}$$

$$\text{Area of the cylinder} = \frac{\text{displacement} \times 144 \times 12}{\text{stroke} \times \text{rev/min} \times 2 \times E_{VS}}$$

$$\text{Area of first stage cylinder} = \frac{612 \times 144 \times 12}{14 \times 327 \times 2 \times 0.769} = 150 \text{ in.}^2$$

First stage diameter $= 14$ in.

$$\text{Area of second stage cylinder} = \frac{288 \times 144 \times 12}{14 \times 327 \times 2 \times 0.762} = 71.4 \text{ in.}^2$$

Second stage diameter $= 9\frac{5}{8}$ in.

The probable machine, therefore, would be a $14 \times 9\frac{5}{8} \times 14$ two-stage compressor driven by a 1200 hp motor. By similar calculation, a four-cylinder machine could be used which would be a $9\frac{3}{4} \times 9\frac{3}{4} \times 6\frac{3}{4} \times 6\frac{3}{4} \times 14$ two-stage compressor at 400 rev/min with a 45,000 lb rod load. Rod load must now be checked from Eq. (8).

14 Cylinder

$$R.L. = P_2 \times A_{HE} - P_1 \times A_{CE}$$
$$= 386 \times 154 - 150 \times 147$$
$$= 37,400 \text{ lb}$$

$9\frac{5}{8}$ Cylinder

$$R.L. = 1000 \times 72.8 - 386 \times 65.8$$
$$= 47,000 \text{ lb}$$

This exceeds the rod load of the two-cylinder machine. Therefore, the four-cylinder machine must be checked for rod load.

9³⁄₄ Cylinder

$$R.L. = 386 \times 74.7 - 150 \times 67.6$$
$$= 18,700 \text{ lb}$$

6³⁄₄ Cylinder

$$R.L. = 1000 \times 35.8 - 386 \times 28.7$$
$$= 24,700 \text{ lb}$$

TABLE 11

Available data on machine	13 in. cylinder	9¾ in. cylinder
Bore, in.	13	9¾
Stroke, in.	16	16
Cylinder design pressure, lb/in.^{2}	900	1,300
Head end clearance, %	18.4	14.8
Crank end clearance, %	17.1	13.8
Suction valve area per end, in.^{2}	8.4	5.3
Discharge valve area per end, in.^{2}	8.4	5.3
Rod size, in.	3¾	3¾
Maximum rod load, lb	75,000	75,000
Speed, rev/min	300	300

This is within the limit of the machine, so the compressor for the application will be a $9\frac{3}{4} \times 9\frac{3}{4} \times 6\frac{3}{4} \times 6\frac{3}{4} \times 14$ two-stage compressor with a 1200 hp, 400 rev/min driver. Note here that the dew point of the gas should be checked at interstage to make certain that no fraction has gone into the two-phase region. Interstage pressure must be altered by cylinder sizing if this occurs.

Existing Machine on New Service. Consider a case where an existing compressor is available which is a $13 \times 9\frac{3}{4} \times 16$ compressor with a 1200 hp driver at 300 rev/min. What will the performance of this machine for the service in Example 1 be? See Table 11.

Conclusion

Horsepower and rod load are acceptable for this machine. Capacity = 5820 × 1440 − 10^6 = 8.4 MM SCFD, which is less than required. Therefore, process capacity requirements could be decreased to accept this capacity or, alternatively, clearance could be reduced to increase E_{VS} to bring capacity closer to process requirements.

Symbols

E_{VS} volumetric efficiency, suction (%)
E_{VD} volumetric efficiency, discharge (%)
C clearance volume (ft^3; m^3)
D piston displacement (ft^3; m^3)
H head (ft · lb/lb$_m$; kN · m/kg)
h enthalpy (Btu/lb; kJ/kg)
k adiabatic exponent (CP/CV; Cp/Cv)
L cylinder loss factor (%)
n polytropic exponent
N moles of gas
P pressure (lb/in.^{2}abs; bar)
Q volume rate of flow (ft^3/min; m^3/h)
R gas constant [$(1544/M_r)$(ft · lb/lb · °F); $(8314/M_r)$(J/kg · °K)]
r pressure ratio (P_2/P_1; P_2/P_1)
T absolute temperature (°Rankine = °F 459.6; °Kelvin = °C + 273.5)
t temperature (°F; °C)
V volume (ft^3; m^3)
v specific volume (ft^3/lb; m^3/kg)
Z compressibility factor

Note: English units are used in reciprocating calculations.

Driver Selection

HEINZ P. BLOCH

First cost economics, space available, operating cost, utilities available, maintenance requirements, physical size, and variable speed capability are only some of the possible factors to be weighed when selecting compressor drivers. Driver selection requires considerable analysis, and the wrong choice can burden a petrochemical facility with reliability concerns and lower profits for years to come. A brief consideration of principal compressor drivers and summary of application ranges (Tables 1 and 2) are followed by a sample economic driver study showing the degree of detail necessary to ensure optimized selection.

Principal compressor drivers can be grouped by the type of energy supplied to the driver and converted to useful shaft work or output torque by the driver. Only three types of input energy are worthy of consideration from practical points of view: Electric power as supplied to motors, combustible fuel (gas, petroleum derivative liquid, coal dust, etc.), and compressed gas or high-pressure vapors (steam or process fluid streams).

Motor Applications

Simplicity and generally troublefree operation are responsible for the popularity of electric motors in compressor drive applications. The term simplicity in this context refers to well-understood construction as well as operation of

TABLE 1   Typical Characteristics of Electric Motors for Large Compressor Drives[a]

	Motor Type	
	Induction	Synchronous
Efficiency, including excitation	0.95	0.97
Power factor	0.90	1.0
Var requirements (for 10,000 hp motor)	3800	0
Starting current, percent of rated	600	400
Torque, percent of full load:		
Starting	70	40
Pull-in	—	60
Breakdown or pull-out	200	150
Service factor	1.0	1.0

[a]Basis: 1.   Ratings ranging from 2,000 to 20,000 hp at 2300 or 4000 V, or 13,200 V.
       2.   Induction 1800 rev/min, synchronous 1200 rev/min with brushless excitation.

these machines. Motors are available in sizes up to approximately 30,000 hp. Although variable speed motors are sometimes used to drive compressors, the overwhelming majority operate at fixed speed and usually require step-up gears to drive the compressor at an economical impeller tip speed.

Major disadvantages of electric motor drives are high energy cost, lack of speed flexibility and thus potentially uneconomic operation in throttled or bypass mode, and dependency upon an energy source over which the user frequently has little or no control.

Electric motors come in all sizes. It has been estimated that over 60% of the drivers in a typical petrochemical plant are comprised of induction motors below 250 hp in size. This size, incidentally, is also the breakpoint for small vs large, or standard vs special motors. Many centrifugal plant air compressors require motor drivers around 800 hp, and the average motor-driven process gas compressor operates at 1000 to 6000 hp. These motors are clearly labeled special equipment, and appropriate design and procurement attention are indicated if troublefree operation is expected.

The squirrel-cage induction motor is the most widely used compressor driver, particularly in the range to 200 hp. Above 1000 or 1250 hp it becomes less economic, although larger sizes are available and used when operational and process reasons dictate.

Squirrel-cage induction motors are simple, rugged, and reliable. They have no rotating windings, slip rings, or commutators, generally associated with other motors. They are utilized with all compressors and may be belted, geared, direct coupled, or flange-mounted. In the latter, the motor stator is fastened to the compressor frame and the rotor is mounted on the compressor shaft. The impeller of small dynamic compressors is often mounted directly on the motor shaft.

The induction motor has reasonable efficiency but its lagging power factor is often a disadvantage, particularly in large sizes. Its current inrush is apt to be quite high.

The synchronous motor is a fixed-speed machine. Some consider it less reliable than the squirrel-cage induction motor because it has a wound rotor to which direct-current excitation must be applied, usually through rotating slip rings. Consequently, a small direct-current supply (motor–generator set or rectifier) is required, together with more elaborate starting equipment. Synchronous motors may be flange-mounted, coupled, geared, or mounted (engine type) on the compressor shaft. The synchronous motor on compressor drive usually has a relatively low current inrush. It frequently is selected for its high or leading power factor.

Reciprocating Engine and Gas Turbine Applications*

Reciprocating engine and gas turbine drivers have little in common except that they directly convert combustible fuels into shaft work. Until a shortage of clean-burning natural gas and sharp increases in energy cost forced a reassessment of long-term economies, gas turbine drivers looked highly attractive for many compressor applications. Reciprocating engines of the gaseous or liquid (diesel) fueled variety are inherently costlier to maintain than

*Text and data following this subheading are derived from *New Compressed Air and Gas Data*, edited by C. W. Gibbs, Ingersoll-Rand Co., Phillipsburg, New Jersey, 1971.

natural gas-fired gas turbines. The efficiency of reciprocating drivers is around 32 to 38%, which is higher than the efficiency of commercially applied simple cycle gas turbines but substantially below the efficiency of combined cycle gas turbines. Reciprocating engine drivers are thus difficult to justify for compressor drive applications.

The selection of an engine is usually based upon project and local economic factors, as well as such specific engine features as cylindering and balance. Four-cycle engines have higher thermal efficiency, consume less lubricating oil, are smoother running, and are easier to start than 2-cycle engines. Two-cycle engines have no inlet valves, may have no exhaust valves, require less displacement, and therefore may have fewer power cylinders. The differences are usually small, particularly in the turbo-charged models. Both types are in extensive use as reciprocating compressor drivers only.

In the heavy-duty class, diesel engines are used rather infrequently for compressor drive. An integral compressor design is available only in the low range of 200 to 1300 hp.

Although the gas turbine as an industrial driver is relatively new, it has been applied to compressors in a large number of applications. It is most suitable for driving such high-speed types as certain helical lobe units and centrifugal or axial-flow dynamic machines. Typical sizes range from 2000 hp to as high as 24,000 hp.

Principal advantages of gas turbine drivers include low installed cost, reliability exceeding that of reciprocating engines, closely matched speed with a variety of dynamic compressors, and good fuel efficiency if the entire system is fully heat-integrated. However, total heat integration may be costly. Cycle considerations are therefore important in gas turbine applications.

The open cycle is a once-through cycle. It employs a compressor (generally of the axial-flow type) to compress air into a combustion chamber or combustor into which a fuel gas or oil is admitted and burned. The resultant high-temperature gas–air mixture flows through an expander turbine. Part of this power is used to drive the axial compressor; the remainder is available as a prime mover.

The simple open cycle is relatively low in thermal efficiency, ranging from 16 to 25%. The principal loss is in the waste of exhaust gas heat.

There are many modifications that may be used to improve the efficiency, the use of a regenerator being the most common. In this configuration the exhaust gas preheats the compressed air entering the combustor, resulting in a lower fuel rate to obtain the same gas temperature to the turbine. An efficiency increase of about 4% may be obtained. An additional refinement is the use of intercooling during compression of the air, thus reducing the power required for a given weight of air.

The gas turbine cycle becomes much more efficient if the hot exhaust gases are used to generate steam, as in a waste heat boiler, or to perform a process function.

Gas turbines are essentially constant-speed units, the efficiency falling rather rapidly when speed is reduced. Single-shaft turbines have an efficient speed range of approximately 10%. When driving a centrifugal compressor, this range will usually be sufficient for control purposes. When greater speed variation is needed, the two-shaft arrangement may be used. With this design the expander is in two sections, one driving the air supply compressor on one shaft and the other having a separate shaft for connection to the loading compressor.

Operating on gas fuel, the gas turbine is about as reliable as a steam turbine, making possible long continuous runs without shutdown for inspection or maintenance. When using distillate or treated residual fuel, the tips of the burner nozzles must be cleaned periodically.

Basically, the open-cycle gas turbine is a very simple machine. However, because of high temperatures, high speeds, and close tolerances, it must have complex apparatus for control, cooling, lubrication, and emergency protection. It must have a sizable starting motor or turbine with a disengaging clutch, or a source of compressed air sufficient for starting purposes. It becomes self-supporting at about 55% speed.

The gas turbine is a lightweight prime mover, capable of rapid start-up and loading, and has no standby losses. It can be arranged to require little or no cooling water, making it advantageous for use in areas where water is scarce. The gas turbine establishes the speed of any driven dynamic compressor unless gears are used.

Efficiency is closely related to air compression ratio and temperature of gas to the turbine. The latter is primarily limited by presently available turbine materials. As better materials become available and cooling methods are improved, gas turbine efficiencies may be increased.

Air compression ratios of 5 to 7 are quite common, and higher ratios have been used for improved efficiency.

The influence of compressor intake temperature and altitude is considerable since they affect the weight of air (and oxygen) available for combustion.

Aircraft jet engines may be adapted to compressor drive by the addition of a power turbine to drive a loading compressor. The result is a two-shaft gas turbine.

Steam Turbine and Gas Expander Applications

Steam Turbines

Direct-connected turbines are readily available to any desired horsepower. The speed of turbines can well match the speed of centrifugal and axial compressors, making direct-drive units common practice. Turbines with speeds to 35,000 rev/min are available. The horsepower available decreases as the speed increases. Steam conditions play a very important role in horsepower versus speed. Condensing turbines with very low exhaust conditions are limited to moderate speeds in the range above 6000 hp. By decreasing the amount of vacuum, higher speeds and horsepowers become readily available.

Direct drive is possible with many helical-lobe rotaries, but reducing gear drive is required with all other rotaries and reciprocating machines. Some centrifugal compressors use speed-up rather than reducing gears to adapt commercially available lower speed turbines to higher speed compressors.

Good commercial gears are available with a single speed-change up to nine or ten to one. Double-gear trains are required for higher ratios.

Steam turbine drivers may be designed for a wide range of steam conditions and are therefore readily adaptable, within economic limits, to whatever steam may be available. Where exhaust steam from the turbine can be utilized in the plant system, a back-pressure turbine may be used. Where steam consumption is of prime importance, a multistage condensing turbine may be selected. When low-pressure steam is available continuously or intermittently to supplement high-pressure steam, a mixed pressure turbine may be selected. Also, an automatic-extraction turbine can be designed to bleed varying amounts of low-pressure steam to a process or other equipment.

Steam Engines

The steam engine was probably the first prime mover to drive a compressor. It is still used to some extent, although electric motors, steam turbines, and internal combustion engines have superseded steam engine power to a large degree. Steam engines are used only for reciprocating compressors.

The steam engine is a very reliable driver. In most designs it is extremely flexible in operation. The economy holds up well at reduced load (speed).

This driver is best applied where (1) it satisfies a plant heat balance need or (2) the flexibility of variable speed control is a decided process advantage. For some processes utilizing exhaust steam, however, the steam engine is ruled out because it introduces cylinder lubricating oil into the exhaust steam. Where oil in steam cannot be tolerated, the less efficient, somewhat less flexible turbine driver must be used.

All steam-engine-driven units are of the continuous heavy duty type. Most are built integrally with the compressor, only a few being coupled.

Expanders

Expanders are used for two purposes. One is pure power recovery, using gas that exists at an elevated pressure but is usable only at a lower pressure. The reduction must be by mechanical means if its energy is to be conserved. The second purpose is to obtain refrigeration or to lower gas temperature by the process of removing work energy mechanically. There are various expanders, the reciprocating and turbine types being most prominent. Helical-lobe rotary compressors are sometimes used in reverse as engines or expanders.

Summary of Driver Characteristics

Table 2 shows drivers generally used with compressors. Wound rotor motors are not mentioned since their possible advantage of variable speed is generally negated by other and more economic methods of capacity control, and there is little power saving at reduced speeds.

Not all possible combinations of compressors and drivers are shown. The table shows the approximate size range and characteristics of the principal drivers for compressors. Characteristics are typical and may not be exact for all sizes of a given driver. The maximum sizes shown are commercial; larger units are often available.

Detailed Economic Driver Studies*

Availability and Costs—General

From the standpoint of operating reliability and availability, the user has several alternatives in choosing the type or the combination of types of drivers selected for compressor drive applications. Steam turbines, gas turbines, and electric motors supplied from a good electric power source all have an excellent availability record when properly applied, designed, installed, and maintained.

Along with the emphasis on availability, there is, and will be, a continuing emphasis on low investment costs. Electric motors—supplied with power from the petrochemical plant's own generating station or from an outside source—

*Text and data following this subheading are based on ASME Paper 71-Pet-32, "Selecting the Economic Driver System for Large Compressors" by W. B. Wilson and W. B. Palmer. The year 1971 is used as the basis for all cost and technical data.

TABLE 2 General Characteristics of Compressor Drivers[a]

Driver	Horsepower Range	Available Speed (rev/min) (60-cycle power)	Possible Speed Variation	Efficiency	Starting Torque and Amperes % Full-Load	Stalling Torque % Full-Load
Induction motor	1 to 5,000 (or larger)	$3,600/N$ less 2%. $N = 1$ through 8[b]	Constant speed	10 hp, 86% 100 hp, 91% 1000 hp, 94%	60 to 100% torque. 550 to 650% amperes	150% min (or more)
Synchronous motor	100 to 20,000 (or larger)	$3,600/N$. $N = 2$ through 20[b]	Constant speed	93 to 97%	40% torque under 514 rev/min. 40 to 100% torque at 514 rev/min and over. 300 to 500% amperes	150%
Steam engine	55 to 4,000	400 to 140	100% down to 20%	50 to 75% RCE[c]	About 120%	About 115%
Steam turbine	To 40,000 (or larger)	34,000 to 1,800	100% down to 25%[d]	35 to 82% RCE[c]	175 to 300%	Up to 300%
Combustion gas turbine	3,000 hp at 10,000 rev/min to 20,000 hp at 3,000 rev/min (1,000 ft altitude and 80°F)		100% down to 55%	16 to 25% overall thermal efficiency for simple open cycle. 27 to 30% with regenerator	Both single- and two-shaft designs require a sizable starting motor or turbine. The single-shaft design has poor part load torque characteristics and requires a larger starter. Two-shaft turbines have good torque characteristics	

Integral gas engine	From 85 hp at 600 rev/min to 5,000 to 6,000 at approx. 300 rev/min	100% down to 60%[d]	Up to 40% overall thermal efficiency	Nil, started with compressed air	About 120%
Integral diesel engine	From 100 hp at 600 rev/min to approx. 1,300 at lower rev/min	100% down to 60%[d]	32% (HHV)[e]	Nil, started with compressed air	About 120%
Coupled gas or diesel engine	From 100 hp at 600 rev/min to 5,000 hp or more at approx. 360 rev/min	100% down to 60%[d]	41% on gas (LHV) and 36.6% on oil (HHV)[e]	Nil, started with compressed air	About 120%

[a]Tabulation is limited to the commercially applied ranges of horsepower and revolutions per minute.
[b]N is number of pairs of poles.
[c]Rankine cycle efficiency.
[d]Depends on design and existence of critical shaft speeds.
[e]High heat value or low heat value.

will usually result in a lower "battery limits" investment than if steam turbines, gas turbines, or combined steam-gas turbine cycles are used for the compressor drives. So look to savings in overall plant energy costs plus benefits such as increased operating flexibility and availability when checking the profitability of any increased plant investment required for turbines instead of motors for large compressor drives.

When purchased electric power is to power motors for the large compressors, the electrical system may extend for many miles, with a corresponding increase in vulnerability to "minor" disturbances. The word "minor" is used here to stress the importance of a clear understanding between the user and the electric utility as to the quality of electrical service required for profitable operation of a large process plant. Expenditures necessary to provide the required quality of electrical service are justifiable, and certainly must not be avoided to save investment costs.

Energy Costs for Compressor Drive

Driver selection studies place heavy emphasis on comparing turbine energy costs with purchased power costs for different turbine cycles. These range from the most efficient noncondensing steam turbine cycle, where all heat in the exhaust steam is required for process, to the lower cycle efficiency condensing steam turbine or simple cycle gas turbine supplying horsepower only.

An energy cost comparison study should attempt to define the "break-even" electric power cost for motors compared to different turbine cycles. The findings of such a study are best displayed in graphic form.

. But relative energy costs alone do not provide a basis for economic selection of alternative systems. They only show the energy component of savings that can be realized to offset higher investment costs. The system with the lowest energy cost is not necessarily the most economic.

Types of Drivers—Their Costs and Characteristics

Motors

Synchronous vs Induction-Characteristics

Standard induction and synchronous motors have different characteristics that must be considered. Typical characteristics for synchronous and induction motors are obtainable from major vendors.

One of the considerations in selecting the type of motor is motor torque capability at reduced voltage. Reduced voltages may occur when another large motor on the system is being started or when there is an electrical system fault or disturbance. Torque capability of the induction motor varies as the square of the voltage, whereas pull-out torque of the synchronous motor with excitation supplied from a constant voltage source varies directly with motor terminal voltage. For the characteristics listed in Table 1, breakdown torque capability of the induction motor would drop to 100% of rated torque when voltage drops to 71% of motor voltage rating. The synchronous motor would maintain 100% pull-out torque down to 67% volts.

If the synchronous motor with standard characteristics has the starting, accelerating, and pull-in torques needed for the application, motor starting can be accomplished with a smaller system voltage drop. For instance, with a unit transformer-motor arrangement, starting a 10,000-hp motor would cause a voltage dip of 29% at transformer secondary (motor terminals) when a 6 × inrush induction motor is started and 19% when a 4 × inrush synchronous motor is started. The voltage dips are based on a 15-mVA transformer (7.7% impedance) supplied from a 13.8-kV power system having 500 mVA short-circuit duty. Voltage dip on the power system side of the transformer would be 10% for the induction and 6% for the synchronous motor.

When the application requires it, increased torques or other characteristics can be provided with either the induction or synchronous motor. If both higher torques and var supply are desired, then consideration should be given to use of a 0.8-PF synchronous motor instead of a 1.0-PF motor.

Synchronous vs Induction—Prices and Voltage Selection

Estimated price factors for 4000-V, totally enclosed air-water-cooled (TEAWC) motors are shown in Fig. 1.

To simplify the comparisons in Fig. 1, 4000 V motors were used for all ratings. In an actual application, the larger motors would not often be supplied direct from a 4160-V system. The economics of 2300 or 4000 V motors with unit transformers and switching on the high side of the transformer would have to be compared with the use of 13,200 V motors. Even though 13.2 kV motors are designed for high reliability, in many process plant environments a lower voltage motor, having equally careful design, should provide an extra margin of reliability. So, when compatible with good electrical system design, consideration should be given to selecting the lower motor voltage.

Data in Fig. 2 show the effect voltage has on the relative costs of different motors and starting systems. All data are based on TEAWC motors. Curves at the lower left in Fig. 2 indicate the cost of 2.3 and 4 kV motors and motor starters. Where compatible with the plant electrical system, these voltages should be favored out to about 5000 hp at 2.3 kV and 10,000 hp at 4 kV.

The 13.2-kV motor will find little application below about 5000 hp because the transformer combination with a 2.3 or 4 kV motor and a 13.8-kV starter has a lower total equipment cost for induction and about the same as for a synchronous at 5000 hp.

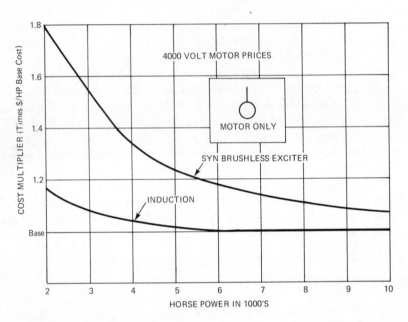

FIG. 1.   Estimated price factors for totally enclosed air/water-cooled motors. Basis: Characteristics as listed in Table 1.

On the basis of overall economics, the unit transformer arrangement (the dashed curves) will be the logical choice for matching driver requirements and electrical system requirements in some process plants.

For motor ratings below about 3000 or 4000 hp, 2300 V motors and starters have a lower price than 4000 V motors. Even so, electrical system considerations may make it more economic to establish a 4.16-kV system to supply a group of motors.

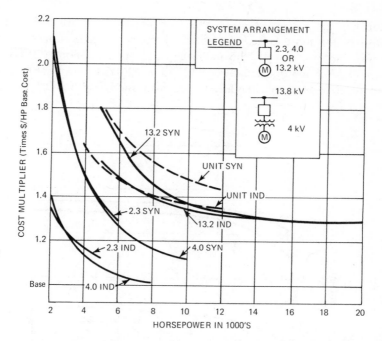

FIG. 2. Effect of motor voltage and power supply arrangement on cost of induction (Ind) and synchronous (Syn) motors. Basis: (1) Solid curves based on power supply at motor voltage (see system arrangement insert). (2) Dashed curves based on 13.8 kV power system, 13.8 kV/4.16 kV unit transformer, and 4 kV motor. (3) Metal-clad starting equipment: (a) 250 mVA for 2.4 kV electrical system, (b) 250 mVA for 4.16 kV electrical system, and (c) 500 mVA for 13.8 kV electrical system. (4) Synchronous motors are 1.0 pF with brushless exciter. (5) All motors totally enclosed air/water-cooled.

Motor Enclosures

With the continuing emphasis on reliability, process plant designers are trending to TEAWC enclosures for motors to power large compressors. A TEAWC motor is a totally enclosed motor which is cooled by circulating air within the motor enclosure. The air is cooled by circulating cooling water in an integrally mounted water-cooled heat exchanger. Experience often indicates this enclosure to be most compatible with management's investment, reliability, and maintenance criteria for large motor selection. The user and/or his process engineering contractor must factor in any special enclosure requirements due to site, atmospheric, and other operating conditions.

From the standpoint of reliability and maintenance costs, the TEAWC motor has an advantage over open, drip-proof, and weather-protected enclosures for most large, high-speed compressor applications. For synchronous motors rated 5000 hp and larger, and induction motors 7000 hp and larger, TEAWC motors actually cost less than WP II as shown in Fig. 3. Even compared to open motors, the TEAWC enclosure adds less than 20% to the cost

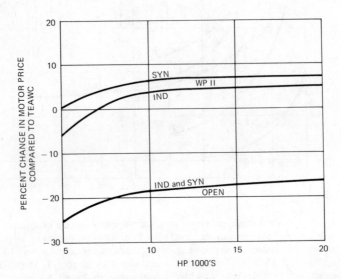

FIG. 3.   Relative costs for open and weather protected II (WPII) motor enclosures versus totally enclosed air/water-cooled enclosures (TEAWC). Basis: (1) Minus means TEAWC cost is more; plus means cost is less. (2) Synchronous is 1.0 PF with brushless exciter.

for ratings above 8000 hp. Cooling water for a 10,000-hp motor will cost approximately $2000/year based on 2¢/1000 gal for 80°F cooling water.

Where site considerations favor a more positive enclosure than the TEAWC, the enclosure is designed for pressurizing with an inert gas, such as CO_2, N_2, or clean dry instrument air. In this case the shaft seals and access covers will be arranged to limit leakage from the enclosure. A 10,000-hp synchronous motor suitable for pressurizing will cost about $7000 more than the TEAWC enclosure. (Cost figures represent approximate values as of 1971.)

Other types of enclosures are available, but they find little application for motors driving the large high-speed compressors in the process industries.

Steam Turbines

Steam turbines, specifically designed for adjustable speed compressor drive, are available in the horsepower rating needed for your applications. These custom-designed turbines can be manufactured from proven component parts for the steam conditions and speeds most economic for overall application.

Some features of the steam turbine that make it an economic driver for large compressors are:

1. Speeds up to 10,000 rev/min or higher are available to match compressor speeds without a stepup gear.
2. Wide adjustable speed range is an advantage when evaluating process operating flexibility and control.
3. They are acceptable for installation in areas with explosive atmospheres.
4. Low fuel-cost power can be produced when all heat in the steam extracted or exhausted from the turbine is needed in process.

The type or types of steam turbines selected for a specific application will depend on the plant's steam balance.

Condensing

The condensing turbine is used for applications requiring shaft power only and where any requirement for process heat is provided from other sources.

Steam for the condensing turbine may be supplied from conventional power boilers, heat-recovery boilers, and/or the extraction or exhaust from other steam turbines. Heat-recovery boilers may recover heat from process or heat from the exhaust of a gas turbine. Turbines for this steam supply may be driving other compressors or, perhaps, an electric generator for in-plant electric power generation.

Even though we normally think of steam turbines having inlet pressures of 400 to 1250 or 1450 lb/in.^{2}gauge, this does not have to be the case. Energy available in expanding steam is the same when 15 lb/in.^{2}gauge–250°F steam is expanded to 4-in.Hg abs as when 250 lb/in.^{2}gauge–500°F steam is expanded to 15 lb/in.^{2}gauge. So when there is the possibility of recovering low-pressure steam from some heat source in process, do not overlook the possibility of using this in a steam turbine. For instance, a successful turbine installation is generating 5400 hp at 5200 rev/min when supplied with 120,000 lb/h of 15 lb/in.^{2}gauge saturated steam.

Noncondensing

When the noncondensing turbine's exhaust steam can be utilized for process heat or steam supply to other turbine drives, its fuel chargeable to power is one-half that of the most efficient turbine installed for power generation only.

Turbine exhaust flow varies as the horsepower requirement of the driven equipment changes. To meet power demand when process steam requirements change and to prevent blowing steam to atmosphere, the overall system must include other equipment that can provide or absorb the difference in steam flow required for process and that exhausted by the turbine.

Automatic Extraction-Admission-Condensing

One or more automatic extraction, automatic admission, or automatic extraction-admission-condensing turbines can be installed to compensate for the swings in process steam use and, at the same time, control the steam pressure in one or more process lines. These turbines automatically extract additional steam when the process requires more than is being supplied by other sources, and admits steam for further expansion in the turbine when there is an excess of steam in the process system. Horsepower output can be maintained constant or varied simultaneously with changes in process steam flow. The turbine's control mechanism automatically varies the turbine inlet and exhaust steam flows as required.

The schematic diagram of Fig. 4 illustrates how variations in steam flow can

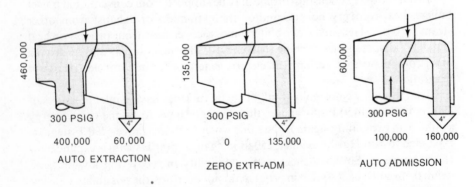

FIG. 4. Schematic diagram indicating alternate steam flow paths in autoextraction-admission-condensing steam turbines. Basis: Initial steam at 850 lb/in.^{2} gauge and 825°F. Turbine output is 20,000 hp.

be accommodated by an automatic admission-extraction-condensing turbine while maintaining constant shaft output. The diagram is based on a 20,000-hp single automatic extraction-admission-condensing turbine with initial steam conditions of 850 lb/in.^{2}gauge–825°F, and automatic extraction-admission at 300 lb/in.^{2}gauge. The turbine governing system automatically controls the 300-lb/in.^{2}gauge process pressure by controlling the steam flow in or out of the turbine at the 300-lb/in.^{2}gauge opening while maintaining the desired power output by varying flow to the throttle and the condenser.

In the extraction mode of operation, 460,000 lb/h of 850 lb/in.^{2}gauge steam enters the turbine with 400,000 lb/h extracted to supply the 300-lb/in.^{2}gauge process requirement and simultaneously generate a large block of by-product horsepower (Fig. 4a).

The zero extraction-admission operation maintains the 20,000-hp turbine output by expanding turbine inlet steam to the 4-in.Hg abs turbine exhaust pressure (Fig. 4b).

During automatic admission there is a surplus of 300 lb/in.^{2}gauge steam from process that enters the turbine at the 300-lb/in.^{2}gauge admission opening. This 100,000 lb/h surplus steam that might have otherwise been wasted to atmosphere generates turbine horsepower as it expands to turbine exhaust pressure. Note how the 850-lb/in.^{2}gauge steam requirement is reduced to 60,000 lb/h (Fig. 4c).

Steam Turbine Performance

The pounds of turbine steam flow required to generate a horsepower-hour depends on many things such as speed, rating, and, of course, steam conditions. For high-speed compressor drives in the 5,000 to 50,000 hp range, turbine efficiency will usually range from 73 to 83%. The major factor in performance is the theoretical energy available as steam is expanded from the turbine inlet conditions to the extraction or exhaust pressure(s) required in the process or to the condenser pressure in a condensing or automatic extraction-condensing turbine.

For the straight condensing or noncondensing turbine, performance can be stated in terms of turbine flow in pounds per horsepower-hour. For an automatic extraction turbine, a performance curve permits the user to determine turbine throttle flow for different horsepower loads and different flows of steam automatically extracted to provide the plant's process heat needs.

Performance of a 24,200-hp, 5,000-rev/min single automatic extraction-condensing steam turbine is shown in Fig. 5. Steam conditions are 850

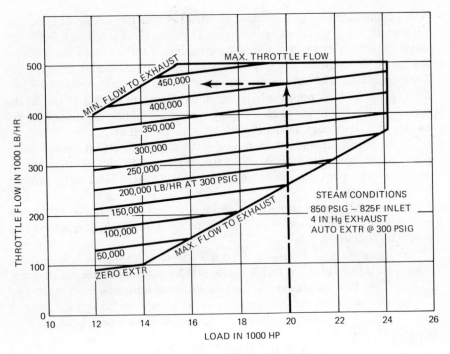

FIG. 5. Steam turbine performance.

lb/in.^{2} gauge–825°F at the turbine inlet with automatic extraction to process at 300 lb/in.^{2} gauge and turbine exhaust to 4 in.Hg abs.

An example will illustrate the use of Fig. 5. Simply enter the curve at the shaft power desired (20,000 hp), then move vertically to the intersection of the line indicating the desired flow of 300 lb/in.^{2} gauge extraction steam (400,000 lb/h), and move horizontally to the left and read 460,000 lb/h turbine throttle flow on the vertical scale. Fuel chargeable to steam turbine power based on 84% boiler and incremental power generation auxiliaries is approximately 6500 Btu/hp·h with this particular "mix" of process steam and horsepower requirement.

By-Product Power

The effect of steam conditions on turbine performance was mentioned earlier. The low fuel cost, by-product power, and the condensing power generated with different turbine inlet steam conditions are shown in Fig. 6. The by-product power generated with 850 lb/in.^{2} gauge–825°F initial steam conditions in-

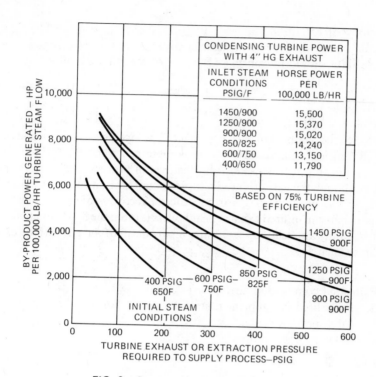

FIG. 6. Steam turbine power generation.

creases from 2650 hp when a process requires 400 lb/in.^{2}gauge steam pressure to 4650 hp if the required process pressure can be reduced to 200 lb/in.^{2}gauge.

If the 400-lb/in.^{2}gauge process pressure cannot be reduced, then the use of 1450 lb/in.^{2}gauge initial steam conditions, instead of 850 lb/in.^{2}gauge, will provide essentially the same savings in purchased energy costs.

The anticipated savings in energy costs for each 100,000 lb/h process steam flow must be evaluated on the basis of different investment costs for the different alternatives. The 200-lb/in.^{2}gauge process steam pressure (if possible) will probably result in higher costs for process equipment than when process steam is supplied at 400 lb/in.^{2}gauge. Likewise, the turbine and boiler system will probably have higher costs when 1450 lb/in.^{2}gauge turbine inlet pressure is selected instead of 850 lb/in.^{2}gauge.

Steam Turbine Prices

The relative prices of condensing and noncondensing turbines with initial steam conditions of 850 lb/in.^{2}gauge–750°F are shown in Fig. 7. The effect of operating speed will depend on steam conditions, as well as horsepower rating, so these data should be considered "order-of-magnitude" only.

Gas Turbines

Wide use of the combustion gas turbine has developed because thermodynamic cycle efficiencies higher than 80% can be obtained when the turbine's useful outputs are carefully integrated with the overall process requirements.

Even though the gas turbine can supply compressed air for process and can utilize surplus pressurized gas from process, recovery of heat from the gas turbine exhaust in unfired exhaust heat boilers is often the first thing that comes to mind. Many applications can utilize a part or all of the gas turbine exhaust for things such as:

1. Preheated combustion air for:
   a. Fully fired or supplementary fired power boilers
   b. Fired process heaters
   c. Catalyst regeneration

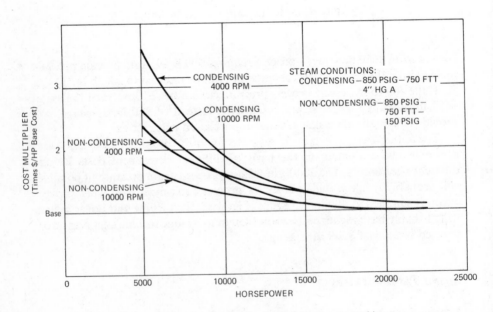

FIG. 7. Estimated prices for mechanical drive steam turbines.

2. Hot gas for:
   a. Exhaust heat-recovery boilers
   b. Heating feedwater for existing boilers
   c. Direct use in process

Obviously, cycle efficiencies are highest when full utilization can be made of both the heat and the oxygen in the gas turbine exhaust, such as in fully fired power boilers or fully fired process heaters.

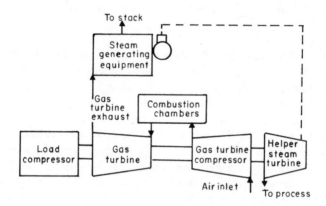

FIG. 8.   Integrated gas turbine–steam turbine cycle.

Gas Turbine Ratings

Even though gas turbines are not available in all the horsepower ratings desired, the use of helper steam turbines can effectively provide an infinite number of ratings. In addition to providing rating flexibility, the helper steam turbine can often exhaust just the steam required for process and/or feedwater heating.

Sea level and 59°F are the basis of the rating adopted by the International Standards Organization (ISO). For 80°F, a sea level site with a 3-in.H_2O inlet drop and a 10-in.H_2O exhaust drop (typical of industrial installations with exhaust heat boiler), the site capability of a 23,350-hp ISO gas turbine would be 20,530 hp. So size your compressor drive system with the site ambient temperature and other site conditions in mind.

Gas Turbine Prices

In process applications the simple cycle gas turbine is normally selected with heat in the exhaust being recovered for some process or power use. Gas turbines have found economic application for large compressor drives in many process plants. The gas turbine is a self-contained prime mover—some call it a complete power plant. Approximate prices for the 25,000 or 32,550 hp ISO rated units are in the $60/hp range. This includes base mounting with their own lubricating oil and governing system.

Steaming Capability and Fuel Chargeable to Power for Specific Gas Turbines

Even with unfired gas turbine exhaust heat-recovery boilers, the ability to generate a sizeable block of low-cost noncondensing steam turbine power usually makes it profitable to select a boiler designed for a pressure higher than the steam pressure required in process. Such a cycle is indicated in Fig. 8.

Typical gas turbine and heat-recovery steam generator (HRSG) performance data are shown in Table 3. The 80°F, sea level site performance is shown for four different gas turbines with ratings ranging from 14,600 to 65,400 hp. These data show steam generating capabilities for gas turbine exhaust heat boilers supplying steam at pressures ranging from 150 to 1525 lb/in.² gauge. For those applications requiring additional power, a part or all of this steam can be fed to steam turbines for expansion to the pressure required in process or to condenser. When the requirement for process steam is low, then the unfired or supplementary fired boiler will often be the economic selection, but do not forget other direct process uses for the high-temperature, oxygen-rich, gas turbine exhaust. A fired process heater or reformer, using gas turbine exhaust, can provide the same opportunity to generate low fuel cost gas turbine power as is realized in exhaust heat-recovery boilers.

Looking at the 25,000-hp gas turbine (Table 2), with 895 lb/in.² gauge–830°F steam generation, an 850 lb/in.² gauge–825°F steam turbine exhausting to a 50-lb/in.² gauge process generates 6,300 hp with an unfired boiler, 12,100 hp with a supplementary fired boiler, and 40,800 hp with a fully fired boiler. This is low fuel cost, by-product steam turbine power. For straight condensing turbines exhausting to 4-in.Hg abs, steam turbine power would be 11,700, 22,400, and 75,500 hp for the unfired, supplementary fired, and fully fired boilers, respectively.

TABLE 3 Steam Generation, Fuel Chargeable to Power, and Steam Turbine Power Available with Gas Turbines and Exhaust Heat Boilers. Natural Gas Fuel

	14,600	25,000	32,550	65,400
Gas turbine rating ISO (hp)	14,600	25,000	32,550	65,400
Site performance at 80 F sea level (unfired/supplementary fired/fully fired):[a]				
Gas turbine:				
hp	12,990/12,850/12,650	21,900/21,670/21,340	28,550/28,200/27,700	57,410/56,820/55,940
Shaft speed (rev/min)	6500	4860	4670	3020
Fuel[b] (million Btu/h)	147.2	247.7	309.1	619.1
Exhaust:				
Flow (1000 lb/h)	389.1	686.2	877.6	1799.8
Temperature (°F)	1011/1014/1018	991/993/996	954/957/961	953/956/960
Gas turbine HRSG Fuel (million Btu/h)	0/47.6/333.9	0/88.1/600	0/123.776	0/252.1606

Steam conditions	Steam generated (1000 lb/h)	Available steam turbine power (hp, 1000's)[c]	Fuel charged to gas turbine power (Btu/hp·h)	Steam generated (1000 lb/h)	Available steam turbine power (hp, 1000's)[c]	Fuel charged to gas turbine power (Btu/hp·h)	Steam generated (1000 lb/h)	Available steam turbine power (hp, 1000's)[c]	Fuel charged to gas turbine power (Btu/hp·h)

TABLE 3 (continued)

Condition	Turbine exhaust				Turbine exhaust				Turbine exhaust				Turbine exhaust			
	4 in. Hg abs		50 lb/in.^{2} gauge		4 in. Hg abs		50 lb/in.^{2} gauge		4 in. Hg abs		50 lb/in.^{2} gauge		4 in. Hg abs		50 lb/in.^{2} gauge	
Unfired boilers																
150 lb/in.^{2} gauge–saturated	67.5	5.8	1.5	5090	115	9.8	2.5	5010	138	11.8	3.0	5030	282	23.8	5.8	4900
420 lb/in.^{2} gauge–655°F	55.6	6.6	2.9	5460	94.2	11.1	4.9	5410	112	13.2	5.8	5460	229	27.0	11.8	5330
630 lb/in.^{2} gauge–755°F	51.7	6.8	3.4	5650	87.4	11.5	5.7	5620	103	13.6	6.7	5680	211	27.7	13.8	5560
895 lb/in.^{2} gauge–830°F	48.8	7.0	3.8	5830	82.1	11.7	6.3	5820	96.1	13.7	7.4	5900	196	28.0	15.1	5780
Supplementary fired																
420 lb/in.^{2} gauge–655°F	96.5	11.4	5.0	4870	170	20.1	8.8	4740	218	25.7	11.3	4730	446	52.6	23.1	4580
630 lb/in.^{2} gauge–755°F	92.1	12.1	6.0	4940	162	21.3	10.6	4820	208	27.4	13.6	4810	426	56.1	27.9	4660
895 lb/in.^{2} gauge–830°F	89.2	12.7	6.9	5010	157	22.4	12.1	4880	201	28.6	15.5	4880	412	58.7	31.7	4730
1315 lb/in.^{2} gauge–905°F	86.5	13.3	7.7	5080	152	23.4	13.6	4970	195	30.0	17.4	4950	400	61.5	35.7	4800
1525 lb/in.^{2} gauge–955°F	84.4	13.5	8.1	5150	149	23.9	14.2	5040	190	30.4	18.2	5020	390	62.5	37.2	4880
Fully fired																
630 lb/in.^{2} gauge–755°F	305.5	40.1	20.0	3670	544	71.6	35.6	3470	694	91.3	45.5	3490	1434	188.7	93.9	3310
895 lb/in.^{2} gauge–830°F	297.5	42.4	22.9	3670	530	75.5	40.8	3470	676	96.3	52.0	3490	1397	199.0	107.5	3310
1315 lb/in.^{2} gauge–905°F	291.0	44.7	25.9	3670	518	79.6	46.2	3470	661	101.6	58.9	3490	1366	210.0	121.8	3310
1525 lb/in.^{2} gauge–955°F	285.5	45.7	27.2	3670	509	81.6	48.6	3470	649	104.0	62.0	3490	1341	214.9	128.1	3310

[a] Gas turbines and boilers fueled with natural gas. Unfired boiler 92% effectiveness for SH and evaporator, supplementary fired units 86.8 to 90.5% effectiveness with 1400°F gas to boiler, fully fired units fired to 10% excess air with 300°F stack. Assume 3% exhaust bypass stack damper leakage, 3% blowdown, 1½% radiation and unaccounted losses, and 228°F feedwater for all cases. Gas turbine inlet 3 in. H_2O, exhaust 10 in. H_2O for unfired, 14 in. H_2O supplementary fired, and 20 in. H_2O for fully fired.

[b] All fuel data based on HHV natural gas. Fuel chargeable to gas turbine power assumes gas turbine credited with incremental power generation auxiliaries and equivalent 84% boiler fuel required to generate steam.

[c] Steam turbine power assumes all boiler steam expanded in 75% efficient turbines to exhaust pressure indicated. Assumes 5% pressure and 5°F temperature drop between boiler outlet and turbine inlet.

Example to Indicate Driver Selection Economics

General and Base Motor Drive Alternative

The economic selection of driver types will depend on the overall process plant requirements for power, process heat, and combustion air, plus the availability of by-product fuels or heat energy available from the process that might be used in a turbine cycle.

The approach to making this selection can best be shown with a *numerical* example problem. Please note that the pricing reflects 1971 cost data and is quite evidently no longer realistic.

Process plant requirements and a system with all electric motor drive for the example to be developed here are shown in Fig. 9. Note that the five large compressors are all driven by electric motors, and low-pressure process boilers provide the plant requirements for process steam. On the basis that a reliable source of purchased electric power is available, this diagram indicates our minimum investment base alternative. Table 4 gives a more detailed listing of process plant requirements and site conditions to be used in our comparison of alternatives.

With the large requirement for process steam, higher pressure boilers with steam turbines to drive the compressors are, perhaps, the first alternative that comes to mind. A quick estimate (using data from Fig. 6) indicates that 850 lb/in.^{2}gauge–825°F steam expanded through steam turbines to supply all the process steam requirements would generate about 28,000 hp, and an additional 39,000 hp would be generated by passing steam to the condenser. This sizeable requirement for condensing power indicates the possible opportunity for a large gas turbine driver (about 30,000 hp) with turbine exhaust used as highly preheated (950°F) combustion air for the 895-lb/in.^{2}gauge boiler.

A higher requirement for process steam would have made more steam turbine power economic. A lower requirement may have favored more gas turbine power with exhaust to unfired or supplementary fired boilers.

To simplify this example, the five large compressors were assumed to operate at 4670 rev/min. This permitted any combination of tandem drives and happened to match the speed of an available gas turbine.

Combined Steam–Gas Turbines for Large Compressor Drive

A combined steam–gas turbine driver system alternative is shown in Fig. 10. With turbines for large compressor drivers, the average (8400 h/yr) purchased

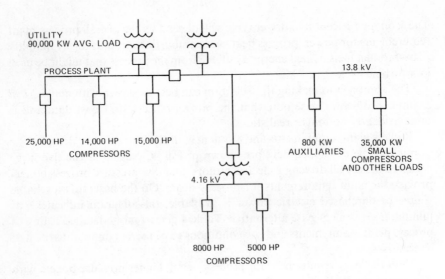

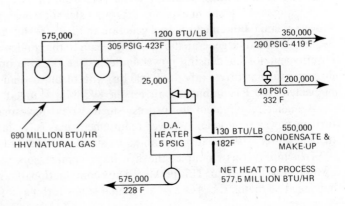

*ALL NUMBERS ARE FLOWS
IN LB/HR UNLESS NOTED OTHERWISE

FIG. 9. Simplified diagram showing a process plant with electric motors driving all compressors. All numbers are flows in lb/h unless otherwise noted.

electric power is reduced from 90,000 to 36,000 kW including the extra power required for boiler feed pumps, plus cooling towers and circulating water pumps for the steam turbine condenser.

This simplified example does not include or show provision for start-up power and/or steam, or any multiple boiler installation that might be considered in an isolated new process plant. But note the operating flexibility that can be provided with the gas turbine, its helper steam turbine, the fuel-fired exhaust heat boiler, and the automatic extraction condensing steam turbine.

Economics

Investment

Now to economics: Estimated investment for 895 lb/in.^{2}gauge boilers, turbines, steam turbines, and cooling towers (Fig. 10), and the base motor drive case (Fig. 9), as well as annual out-of-pocket operating costs for these two alternatives, are summarized in Table 5. These data indicate relative investments (total installed costs) for that part of the process plant system which provides power for the five large compressors and for boilers to supply process heat.

Investment for the turbine drive system is $1,300,000 higher than the motor drive system ($9,100,000 vs $7,800,000), but annual operating costs are $2,485,000 lower ($5,800,000 vs $8,285,000). These savings result in a gross payout of 0.52 years or a gross return of 190% on the extra investment for the turbine drive system. Cumulative net cash flow, discounted rate of return, and present worth are more sophisticated yardsticks of measuring profitability of an increased investment, and these data will be developed next.

Cumulative Net Cash Flow

Figure 11 shows cumulative net cash flow based on purchased fuel and power costs used in our example and indicates the effect of changes in energy costs, and the lower curve shows the effect of 7% state income tax.

Profitability of this $1,300,000 investment—for turbines instead of electric motors for large compressor drives—is based on borrowed capital with the principal being paid off with annual net cash flows during early years of the project. Outstanding debt is charged, and positive cumulative net cash flows are credited with interest at 8%. After repayment of all debt obligations and 48% federal income tax, the extra investment for turbines is recovered in just over 1

TABLE 4 Plant Requirements and Site Data Used in Selecting Economic Energy
          Source

Large compressors (5 units require 55,000 kW purchased electric power if motor drive)
  Horsepower: 25,000, 15,000, 14,000, 8,000 and 5,000 hp at 4670 rev/min
Other small compressors and plant electrical load equivalent to 35,000 kW electric
Heat to process in-process steam:
  Net process steam out − Btu in returns = 577.5 million Btu/h
  350,000 lb/h at 290 lb/in.^{2}gauge–419°F
  200,000 lb/h at 40 lb/in.^{2}gauge–332°F
  Process condensate and make-up returns at 182°F (h = 150)
Purchased energy:
  Natural gas at 200 lb/in.^{2}gauge, 30¢/million Btu HHV
  Electric power:
    7.8 mils/kWh with all motor drive
    8.0 mils/kWh with turbine drives for major compressors
Assumptions and basis of estimating operating costs
  Operation, 8400 h/yr at normal loads listed above
  Gas turbine normal operation at 80°F sea level site, 3 in. H_2O inlet and 20 in. H_2O
    exhaust pressure drop and 3% exhaust bypass damper leakage
  Boilers:
    Ambient boilers (base case), 84% efficient HHV and with 0.3% heat loss before
      process
    Zero blowdown
  Turbines:
    Boiler to turbine, $\Delta P = 5\%$ and $\Delta T = 5°F$
    Turbine to process, $\Delta P = 10$ lb/in.^{2}
  Cooling towers:
    70°F WB, 90°F DB with 20°F approach and 25°F cooling range
  Water, cost in ¢/1000 gal:
    Cooling tower make-up, 2¢
    Boiler feedwater, 80¢

---

year (Curve A). Net positive cash generated is more than $14,000,000 at Year 10
and more than $26,000,000 at Year 16.

If we assume a 50% increase in both fuel and purchased power costs (Curve
B), net cash flow is $42,000,000 at Year 16. And, if we assume power costs stay
constant and fuel costs are increased 100% (Curve C), cumulative net cash flow
is $17,400,000 at Year 16. For those regions where a corporate income tax is
levied, the dashed Curve C indicates the effect state income tax has on net cash
flow compared to Curve C. This dashed curve assumes state income tax is 7%
based on the same taxable income used for federal income tax. This extra tax

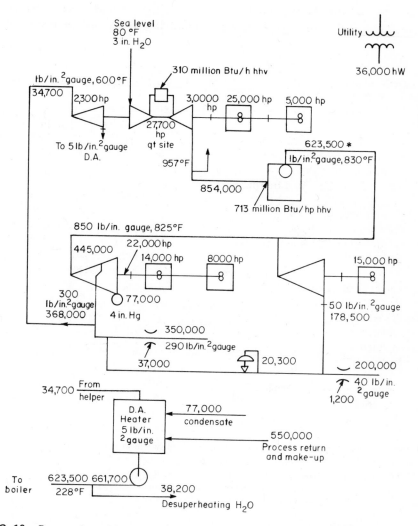

FIG. 10. Process plant with turbines driving large compressors. All numbers are flows in lb/h unless otherwise noted.

reduces positive net cash flow from $9,000,000 to $7,600,000 at the end of Year 10 and from $17,400,000 to $14,400,000 at Year 16.

These curves pinpoint the marked effect increases in purchased power costs have to increase the profitability and the relatively small effect increases in purchased fuel costs have to reduce profitability of the turbine installation.

TABLE 5 Relative Investment and Operating Cost When Large Compressors are Motor Driven Versus Turbine Driven

Driver System for Large Compressors	Motor Drives[a] (Fig. 9)	Turbine Drives (Fig. 10)
Investment costs, $1000		
Motors, switchgear and transformers	1900	
Boilers	5900	5000
Steam turbines		1200
Gas turbine with helper		2700
Condenser and cooling tower		200
Total	7800	9100
Added cost for turbine drive system	$1300	
Annual operating costs, $1000[b]		
Fuel	1740	2550
Power	5900	2420
Labor	340	490
Maintenance	195	228
Water	110	112
Total	8285	5800
Decreased cost for turbine drive system	$2485	
Gross payout for turbine driver	0.52 years	

[a]Base minimum investment case includes total installed system associated with large compressor drives and process boilers only. Approximate pricing for 1971.
[b]Operating costs include fuel and power for entire plant. Other costs associated with turbine drivers, boilers, and associated equipment only. Approximate costs for 1971.

Looking at Year 10: If 50% higher power costs had been assumed in the example (Curve D), the net cash flow resulting from turbine drivers would have been $10.8 million higher ($25.1 million vs $14.3 million). The decrease in net cash flow, if we had assumed fuel 50% higher, would have been only $2.7 million (one-quarter the dollar effect power cost had). Fuel costing 100% more would reduce net cash only $5.3 million compared to the base case at Year 10.

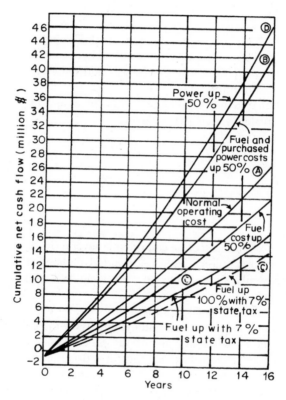

FIG. 11. Cumulative net cash flow for increased investment in turbine drives. Basis: (1) Sum-of-the-years digits depreciation, 16 yr. (2) Local property taxes and insurance, 2.5% of investment. (3) Federal income tax, 48%. (4) Interest paid or credited at 8%. (5) Added initial investment for turbines, $1,300,000. (6) Savings in out-of-pocket operating costs, $2,485,000/yr.

Discounted Rate of Return and Present Worth

The annual net cash flows that result from the turbine installation during a 16-year period were used to develop the data in Table 6.

For this profitable investment the discounted rate of return after all debt obligation is 103% for the plant with present fuel and purchased power costs. If fuel costs increase 50%, DRR only drops 16% to 87%. If purchased power costs increase 50%, the DRR goes up 50% to 173%.

With annual net cash flows discounted at 20%, the present worth is $3,500,000 after all debt obligation and with fuel prices 100% higher than now estimated. This is 2.7 times the $1,300,000 added investment required for turbines instead of motors.

If both purchased power and fuel costs increase 50%, the DRR on Δ investment for turbines increases from 103 to 157%.

Conclusion

The approach used in this section can provide several of the inputs needed for overall driver system studies. Performance and updated price data must be combined with similar data on related plant facilities both inside and outside the battery limits.

Most often, process plant operating costs will be lowest when different types of drivers are combined in the overall system. With the equipment available today, the optimum driver system for the process plant's large high-speed compressors will usually include steam turbine drives and gas turbine drives. And in areas where a reliable source of low cost purchased power is available and there is little need for process heat, electric motors will be economic for some or all the drivers.

Optimum overall plant profits will result when the process plant designer has time to carefully coordinate the capabilities of different type drivers. This will permit supplying the many needs of the overall process plant and, at the same time, make economic utilization of different forms of energy that may be available as a by-product in other process plant areas. The overall look should not be forgotten.

Figure 12 summarizes the use of various drivers in the process industries.

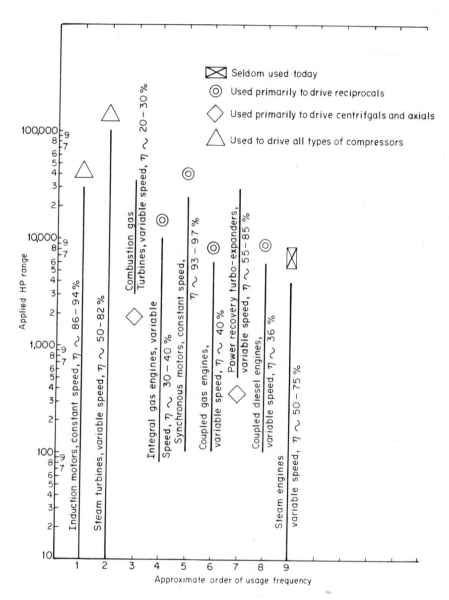

FIG. 12. Drivers used for process plant compressors.

TABLE 6 Discounted Rate of Return and Present Worth of Earnings Realized When an Increased Investment is Made for Turbines to Drive Large Compressors[a]

	Discounted Rate of Return (%)	Present Worth of Earnings, $ Million, Discount Rate				
		5%	8%	12%	16%	20%
Turbine vs base case, normal operating cost	103	16.9	13.2	9.8	7.5	5.9
Fuel up 50%	87	13.9	10.8	8.0	6.1	4.7
Fuel up 100%	71	10.9	8.4	6.2	4.6	3.5
Purchased power up 50%	173	29.9	23.6	17.8	13.8	11.0
Purchased power and fuel each up 50%	157	26.8	21.2	15.9	12.3	9.8

[a]Basis: 16 year life. Net cash flows from Fig. 11.

Installation, Operation, and Maintenance

RALPH JAMES, Jr.

General

The first requirement for a compressor installation is the need to operate in a safe manner. A good running machine of any kind is no accident. Successful (safe, economical) machinery operation is possible only if all four of the following conditions are met:

1. *The design must be right.* The machine must be capable of pumping the required quantity of gas over the range of suction and discharge conditions without overloading either the driver or the compressor end.
2. *Correct materials* must be *used* in order to realize success from a good design.
3. *Assembly* in the shop and in the field must be completed *without error.*
4. The compressor must be *maintained properly.* "Maintenance" is the responsibility of both the process operators and the mechanics. For example, the process operator is exercising poor maintenance if he operates

the machine outside its design limits or neglects lubrication. Likewise, the mechanic is not practicing good maintenance if he "fixes" a machine by the trial and error approach.

Obviously, the starting point for a compressor installation is the design phase which includes the whole system—foundation, physical arrangement, controls, piping, monitoring, and maintenance requirements including spare parts. The selection of compressor type and evaluation analysis should take into account equipment size and cost, thermal efficiency, the service it will perform with special consideration to critical or hazardous applications, and performance requirements including significance of downtime, present and future systems and plant requirements, and the anticipated plant life. The machinery specialist should be involved starting at the predesign consultation stage. Continuity of machinery technology input should cover: (1) mechanical specifications including on-stream monitoring requirements, (2) bidder selection and evaluation, (3) nonprocess quality control during fabrication, (4) shop tests, (5) installation check out, (6) field performance tests, (7) establishing operating guidelines, and (8) setting up maintenance procedures.

During the design and nonprocess quality control stages the maximum use should be made of computerized simulations of system dynamics including rotor dynamics, system dynamics, foundation response, and piping flexibility and dynamics.

Good Maintenance in a "Nutshell"

Two rules define good maintenance practice. These are:

1. If a machine is running well by analytical check, leave it alone.
2. If a machine is not running well, fix only what needs to be fixed.

There is more to good operation than these simple maintenance rules, of course.

The Elements of "Maintenance"

In the optimum case, maintenance involves:

1. *Evaluation* of the condition of a running machine to determine what needs to be done, if anything, for the machine to *safely* and *economically* continue to operate for a prescribed period of time subject to the following restrictions:
   a. Certain material weaknesses (such as fatigue cracks) or previous faulty assembly may not be evident from presently available on-stream analytical techniques.
   b. Some types of impending failure signs may develop and run to completion so rapidly that failure will occur between equipment checks. For example, if the lubricant flow to a heavily loaded antifriction bearing falls below the required amount, the bearing may be completely destroyed in less than a minute.

These restrictions lead to some important special requirements for maintenance. Restriction "a" implies that all material and workmanship must be checked and double checked on critical machinery. Restriction "b" leads to the necessity for continuous monitoring of critical process machinery to provide alarms and/or automatic shutdown devices to minimize the possibility of catastrophic failures.

2. *Determination* of *corrective* measures necessary to return a machine which is down to safe and economical operation. In this case the methods of analysis may differ from those employed for on-stream analysis.
3. *Development* of efficient *work execution procedures* and tools is necessary if the desired level of maintenance is to be provided as inexpensively as possible. Included in this area are:
   a. Proper scheduling of analyzer checks.
   b. Tune-ups.
   c. Improved materials.
   d. Planned, streamlined procedures to minimize oversights or errors in assemblies.
   e. Specialized tools.
   f. Controlled spare parts inventories.
   g. Maximum standardization and interchangeability of parts.
4. *Safety* of personnel is the number one consideration in maintenance and operation as in any other field of endeavor. General guidelines to safety are collected in the section entitled "Compression Equipment Safe Operating Procedures."

Predictive Maintenance vs Periodic Inspection Maintenance

The term "predictive maintenance" is used to identify the maintenance philosophy described above. To sum up this philosophy in slightly different words, predictive maintenance evolved from two basic facts:

1. Certain vital parts last longer and operate better if not frequently taken apart.
2. Operation until complete destruction is not only foolish but costly.

The above facts represent the extremes of the situation. From an economic viewpoint, it is definitely poor policy to be constantly tearing an engine down for inspection. On the other hand, for both safety and economic reasons, we cannot go to complete destruction before exercising some form of maintenance or adjustment. Inasmuch as maintenance is the most important factor in successful operation, an answer has to be found. This answer lies in the fact that thousands upon thousands of operating hours have proven that 99% of all failures are preceded by certain signs, conditions, or indications that these parts were going to fail. This being true, it would be ideal to use these indicators or signs in determining just when an engine should be overhauled. They could also be used to ward off serious premature casualties subject to the limitations mentioned above.

If signs and indicators could be used to determine when an engine is to be overhauled and to ward off serious casualties, then there would not be any forced interruptions of service that are common in the other types of maintenances; that is, it would not be necessary to dismantle the engines for inspections. After all, engines are bought to run, not to be torn down and inspected. The more an engine is on the line, the more horsepower hours are realized from it. Likewise, by operating on the above principle, one would realize the maximum life of all parts. Another important advantage is that the entire personnel of plants will become more familiar with the "know-how" of engines. This fact plus preliminary information on indicators will be borne out later.

For a comparison, let us review the principles of *periodic inspection maintenance*. In this type of maintenance the assemblies constituting the entire engine are torn down, inspected, and cleaned after they have run a specified number of hours. The number of hours between inspections is established by the engine manufacturer on the basis of experimental and development tests, or they are established by the experience of the operators. Although it has its advantages and is used extensively by some competent operators, it has many drawbacks. Needless disassembly of the vital parts of the engine is one big disadvantage. Removal and replacement of parts still in excellent condition, no matter how carefully done, frequently induces trouble. Perhaps this is caused by simply disturbing parts which have found their optimum running fit or finish.

A typical example of this is bearings where the chances of mistakes during assembly and disassembly are excellent. It is very easy to enter dowels improperly. Connecting rod and main bearing bolts are often left loose or they are tightened to such an extent that they distort the caps even when torque wrenches are used. Nicks and burrs on babbitt bearing faces and on journals are very common. Studs, nuts, and bolts tend to lose their effectiveness with constant tearing down. Once the cylinder heads have been removed, it would be unwise to put them back without grinding valves and valve inserts. As a result, this reduces their life due to the increased number of times they have been serviced.

It is claimed by all bearing manufacturers that nine out of ten bearing failures are caused by dirt. A certain amount of dirt gets into the engine during assembly in spite of all precautions and extensive cleaning procedures. It is a constant battle for engine builders to eliminate this dirt. The same problem confronts the operators when they tear down engines for overhaul. Dirt is present even in the cleanest installations. Therefore, the oftener engines are needlessly torn down, the better the chances of introducing dirt. Also, in any engine operation there is always an accumulation of carbon and dirt which collects in places where it is not doing any harm. However, during reassembly, this harmless dirt and carbon may be moved to locations where it can cause damage.

As mentioned previously, the inspection periods are spaced according to past performance of the vital assemblies. However, some small part deep down in the assembly might fail due to material weaknesses or previous faulty assembly. Failure of this small part leads to premature casualty for the entire assembly, so in this type of maintenance the operator is not free from disastrous casualties because, regardless of the amount of experience both the engine builder and operator have had with the engine, no one can predict how long a particular assembly is going to last. In the case of piston rings, it is not uncommon for engines that have operated in the same plant under identical conditions to have one engine give a ring life of 8,000 h whereas the engine right beside it will run for 20,000 h before requiring a ring job. Therefore, to be safe one might establish 8,000 h for the life of the rings. In this case, the engine that ran well to 20,000 h is penalized. On the other hand, if the engine builder or the operator cares to take a chance, he might establish, on the basis of these two engines, an inspection period of 12,000 h. Where the rings only lasted 8,000 h, the engine would have been operated for 4,000 h under poor operating conditions. In the case of the engine that went 20,000 h, the operators would lose 8,000 h of good operation. Due to the inability of predicting the exact life of parts, the operators cannot realize the maximum life of vital assemblies.

One of the main disadvantages of periodic inspection maintenance is that it does not afford a means of training all personnel in the "know-how" of engines. The mechanics are merely machines tearing down the engines and putting them back together. Inspections are carried on by the supervisor who makes any necessary decisions. Therefore, the purpose of the parts and how they function are not properly understood by all personnel.

Another disadvantage is the lack of flexibility. The engine builder cannot adopt any one standard and recommend it to customers who operate engines under various conditions and applications. As a result, the engine builder, of all people, is left without any plan of maintenance.

Responsibility for Predictive Maintenance

To a large extent, it will be the responsibility of the Mechanical Department to provide the desired level of maintenance as inexpensively as possible through the proper scheduling of analyzer checks, tune-ups, improved materials, streamlining procedures, specialized tools, controlled spare parts inventories, maximum standardization, and interchangeability of parts. If the basic equipment owner wants to change the level of maintenance (the service factor will be the best yardstick of this level), the Mechanical Department should be in a position to tell him how much more it will cost for an increase or how much can be saved by a decrease. Cooperation will be necessary in scheduling equipment downtime, handling emergency repairs, and adhering as closely as possible to the priority list established for the purpose of getting the most important job done first.

By having these preplanned procedures scheduled and coordinated with the various equipment owners and executed by the Mechanical Department, faster repairs can be expected. Only work indicated by analyzer checks will be performed and the general level of machine availability will be appreciably raised. This program will provide safer operating equipment and improved service factor at a lower overall cost than a repair-after-breakdown policy.

Functions of the Mechanical Technical Service Section

Mechanical Technical Service plant engineers shall:

1. Assist in establishing and maintaining complete maintenance and cost records for all gas engines and compressors.
2. Review preventive maintenance procedures and intervals between procedures for each gas engine and compressor. Recommend changes in procedures and intervals as required to reduce maintenance costs and to improve service factor.
3. Evaluate new materials and recommend material changes where justified.
4. Investigate special or recurring maintenance problems.
5. Serve as contact man for original equipment manufacturers and suppliers. Keep informed of new developments by these organizations.
6. Serve as contact man for similar groups, thus providing a central point or clearinghouse of information relating to field experience with all gas engines and compressors.
7. In order to provide service to the most important equipment at times when a choice must be made, a priority list must be developed and kept current. This list should be based on the economics of downtime of each compressor or engine. The relative values of equipment at the various locations should be determined by basic equipment owners, the Technical Division, and coordination personnel. It will be the responsibility of the maintenance group to use this list as a guide in planning and scheduling work and in dispatching or shifting repair teams.
8. Assist material control and operations in establishing optimum spare parts stock.
9. Review maintenance costs and service factors, and make recommendations to the basic equipment owner for repair or replacement where justified.

Work Execution

General

1. A periodic check of each engine and compressor of 200 h or more will be made by a two-man analyzer team. The analyzer can be attached to the engine and data obtained without shutting down the engine providing the power and compressor cylinders have indicator valves. Some adjustments

can be made with the engine in operation (such as adjusting the fuel supply to the individual cylinders to balance firing pressures). Other adjustments (such as adjusting timing, changing spark plugs or fuel valves) require shutting the engine down for a short period. The analyzer team will make any "on-stream" adjustments indicated by the analyzer after notifying the chief operator on the unit what is planned. If it is possible to shut the engine down for a short time, the analyzer team will also make minor replacements and adjustments.

2. An analyzer team will be available "on call" at any time trouble on an engine is suspected. It is necessary for the engine to be running for the analyzer to obtain data. There should be enough time available to make the troubleshooting calls and still make a routine check on each engine at the prescribed interval.

3. When the analysis is completed, the team will make a report of its findings, a summary of the condition of the engine and compressor, repairs made by the team, and recommended additional repairs. This report will be sent to the operating department head and supervisor.

4. A trained crew can be expected to analyze and make minor adjustments on two and sometimes three engines in a day. Once the engines are put in good shape, each engine should be checked once in 4 to 6 weeks on a schedule, and there will still be time for emergency analyses.

5. Preventive Maintenance Procedures, General

    a. The analyzer team check will be performed periodically. The team check list covers in detail what will be done on each check. Large or critical engines should be checked once a month; a 2- to 3-month interval on the smaller or less critical engines should be adequate.

    b. Major machinery analytical procedures should be done at a 2-year scheduled interval. A check of bearings and clearances on this schedule is necessary to prevent major repair work.

    c. Major overhauls should be scheduled as indicated by the analyzer checks, and the results of the analyzer findings should be planned to cover only the necessary repairs indicated. Analyzer records and bearing checks should adequately evaluate the general condition of the engine and predict both the proper time and extent of major repairs required. The major overhauls may be as long as 5 or 6 years apart.

Procedures

Typical procedures for integral gas engine-compressors follow. Similar procedures should be developed for each type of major machinery.

Analyzer Team Checklist

This procedure is a minor tune-up. The purpose is to check safety devices, spark plugs, and fuel valves; to adjust timing and load balance; and to determine the general condition of the engine by the use of engine analyzers.

A. Before Engine Is Shut Down

The intent of this preliminary inspection is to locate apparent engine malfunctions before an engine is shut down so that any necessary adjustments, repairs, or replacements can be made in the most efficient manner after an engine is down. All observed malfunctions should be listed, marked, or tagged for easy identification after an engine is shut down.

1. Check the following pressures and temperatures and compare with Mechanical Technical Section or manufacturer's recommendations.
   a. Engine fuel gas pressure.
   b. Oil pressure to bearings.
   c. Oil temperatures in and out of engine.
   d. Water temperatures in and out of engine.
   e. Measure scavenging air pressure on two-cycle engines with a manometer. Abnormal pressure may indicate defective scavenging valves, rings, port fouling, filter fouling, cooler fouling, or turbocharger malfunction. Check turbochargers for correct operating speed, oil pressure, excessive vibration, proper discharge pressure, sticking bypass valve, gasket leaks, and unusual noises.
2. Check for leaking:
   a. Starting air check valves.
   b. Fuel injection valves.
   c. Cylinder head gaskets.
   d. Water pump packing.
3. Check for correct ignition timing and ignition system components.
4. Check for ignition wiring defects.
5. Check all force-feed lubricator pumps for proper operation and adjust rates if necessary.
6. Check for oil leaks at locations such as inspection plates.
7. Visually check external bolts for tightness.
8. Check governor linkage for excessive wear.

9. Take power cylinder compression pressures to detect bad valves and/or faulty piston rings.
10. On four-cycle engines, check fuel mixing valve for free movement.
11. Check overspeed trip and setting when shutting down engine.

B. If Engine Can Be Shut Down

1. Replace defective spark plugs, other ignition components, and defective fuel valves.
2. Visually check explosion relief valve (spring, guide, seat, and discharge). Repair or replace if necessary.
3. Crank engine with air to check operation of starting air valves. Free any sticking valves.
4. Where V-belt drives are used, check V-belts for proper tension and alignment.
5. Check low oil-pressure trip, high water-temperature trip, and automatic fuel gas shutoff valve for proper operation.

C. After Engine Is Started

1. Adjust fuel injection valve tappet clearance.
2. Recheck ignition timing if timing adjustment was made.
3. Adjust governor full-load speed.
4. Balance load between power cylinders.
5. Check and record the following:
   a. Compression pressure.
   b. Firing pressure.
   c. Cylinder exhaust temperatures on two-cycle engines.
   d. Engine speed.
   e. Turbocharger speed where applicable.
   f. (1) Scavenging air intake and discharge pressure on two-cycle engines. Compare with standard data sheet and make necessary corrections. (2) Maximum no-load manifold vacuum (with compressor bypass open) on four-cycle engines. This is peak vacuum obtained by adjusting fuel mixture. Compare with standard data sheet and make necessary corrections.

D.   Compressor

Make a time–pressure diagram on each end of a compressor cylinder where test valves are installed.

    If an engine cannot be shut down while the analyzer team is available, a list of recommended repairs and adjustments should be left at the control room with the chief operator and a copy sent to the operating supervisor.

    Repairs and adjustments made by the analyzer team will be listed; those recommended but not done due to lack of time or any other reason will also be listed.

Procedure GET-2

Procedure GET-2 is a major tune-up and bearing check. It includes the work recommended by the analyzer team as well as a check of main and connecting rod-bearing clearances and crankshaft web deflections. The object is to find and correct any unusual wear or misalignment before a serious failure occurs, as well as to benefit from better performance after the major tune-up.

A.   Before Engine Is Shut Down

Perform items recommended in Section A of "Analyzer Team Checklist."

B.   After Engine Is Shut Down

1.   Perform items recommended in Section B of "Analyzer Team Checklist."
2.   Change oil and clean crankcase. Inspect crankcase for babbitt, brass, or iron particles which may indicate bearing or sprocket wear. If particles are found, locate source and make necessary corrections.
3.   Check connecting rod-bearing clearances by "jacking." Adjust if necessary and record clearances.
4.   Check main bearing clearances with feeler gauge or by "jacking." Adjust if necessary and record clearances.
5.   Check crankshaft web deflections. If web deflection exceeds 0.002 in., appropriate corrective action should be taken. Record measurements.

GAS ENGINE
ANALYZER DATA SHEET

Unit _____
Mfg. & Model _____
Engine No. _____
Rated Horsepower _____
Firing Order _____
Rated Speed _____ Rpm
Speed Before _____ Rpm

Fuel Flow Rate (SCFM) _____
Test Horsepower _____
(Analyzer Check)
Fuel Pressure _____
Timing Specification _____ °BTDC
No. 1 Mag. _____
No. 2 Mag. _____
Speed After _____ Rpm

ANALYSIS

CYL. No.	Firing Peak Psi	Comp. Peak Psi	Ex-Haust	In-Take	Cyl. Exh. Temp	Wrist Pins	Rings	Spark Plugs	Gas Valves	Push Rods	Rock Arm
1											
2											
3											
4											
5											
6											
7											
8											
9											
10											
11											
12											

Comments & Work Done:
_____
_____
_____
_____
_____

(OK)   - Everything "Go" Until Next Analysis.
(NOK-1)- Not OK-1, Schedule Work For Next Shutdown.
(NOK-2)- Not OK-2, Work Should Be Done Now.

Date _____
Crankcase Level _____
Slo-Flow Meter Reading (Gal) _____

	Normal	OK	NOK
Oil Filter D/P Psi			
Oil Filter Out Psi (Oil To Engine)			
Oil Inlet Temp °F			
Oil Out Temp °F			
Lubricator Pumps			
Governor			
Tachometer			
Trabon			
Ignition Wiring			
Coils			
Air Check Valves			
Scav. Air In Psi			
Vacuum At Air Fil.			
Scav. Air Temp.			
Overspeed			
Oil Press. Shut Dn.			
Water Temp. Shut Dn.			
Turbochg. Rpm			
Foundation			
Scav. Air Valves			
Manifold Pressure (Four Cycle)			
Approximate Loading			%

By: _____

FIG. 1

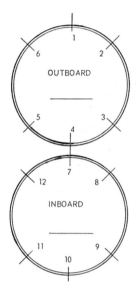

DATE _____

UNIT _____

COMPR. NO. _____

CYLINDER NO. _____

SERVICE _____

BORE & STROKE _____

SUCTION PRESS. _____

SUCTION TEMP. _____

DISCHG. PRESS _____

DISCHG. TEMP. _____

RING MATERIAL _____

RING CONDITION _____

PACKING CONDITION _____

VALVE NO.	TEMP.	LIFT	CONDITION		REMARKS
1					
2					
3					
4					
5					
6					
7					
8					
9					
10					
11					
12					

FIG. 2

6. Check timing chain for excessive slack.
7. Inspect magneto drive shaft bearing; replace if necessary.
8. Clean force-feed lubricator reservoirs.
9. Check flywheel bolts for tightness.
10. Inspect check valves in lubricator lines. Clean or replace if necessary.
11. Calibrate exhaust gas Alnor thermocouple instruments.
12. On Clark TRA and Worthington SUTC engines, inspect scavenging air cooler for broken supports, water leaks, and fouling. On SUTC engines, inspect power rings, pistons, and cylinder bore condition by looking through scavenging ports while scavenging air inspection plates are removed.
13. On Clark TRA-8 engines, open and inspect condition of air blower drive mechanism and bearings. Adjust, repair, or replace defective components.
14. On Worthington SUTC and Clark TLA-8 engines, clean and lubricate pulse generators and check breaker points.
15. Repair or replace all defective manometers, tachometers, thermometers, and pressure gauges.
16. Flush governor and refill with Teresstic 52.
17. Clean scavenging air filter and fill with fresh oil.

C. After Engine Is Started

Request analyzer team to perform items listed in Section C of "Analyzer Team Checklist."

Reciprocating Compressor Maintenance Procedures (CI) Compressor End—Integral Type Drive (Gas, Electric, and Steam Engines)

Procedure CI-1

Clean oil bath air filter and fill with fresh oil (air compressors).

Procedure CI-2

Clean steel mesh-type filters and recoat with oil. Replace felt or fiberglass-type filter elements (air compressors).

Procedure CI-3

Inspect wedge bolts on main and connecting rod bearings and crosshead bushings. Tighten if loose.

Procedure CI-4

Dismantle unloader mechanism, inspect and recondition as required. Clean seats. Clean filter in air line to diaphragms. Check transmitter for unloader.

Procedure CI-5

Change out compressor valves using the following procedure:

1. Remove all valves from cylinder.
2. Clean valve port gasket surfaces. Install new gaskets.
3. Install complete set of reconditioned valves.
4. Where spring-loaded valves are installed, check seating surfaces on both valves and cylinder parts.

Procedure CI-6

Procedure CI-6 consists of a bearing inspection, alignment check, and lubrication system inspection.

1. Inspect for proper clearances the connecting rod bearings, crosshead pin bushings, and crosshead slippers. Adjust or replace if necessary.
2. Inspect main bearings on electric motor drives for proper clearance.
3. Check compressor cylinder support jack screws for tightness.
4. On steam and electric machines, remove force-feed lubricators and install reconditioned lubricators. On all machines, inspect check valves in lubrication lines. (NOTE: Lubricators on gas engine-driven machines will be changed as part of the engine maintenance procedures).

Procedure CI-7

Procedure CI-7 is a compressor cylinder, piston ring, and valve inspection.

1. Pull compressor pistons. Install new piston rings if necessary.
2. Measure pistons and cylinders. Record measurements. Rebore or reline cylinder if measurements exceed limits shown in *Manual of Standard Practices*.
3. Inspect piston rod pressure packing and oil wiper packing. Replace if necessary.
4. Inspect oil lines inside compressor cylinder. Replace as necessary with stainless steel piping.
5. Remove compressor valves. Visually inspect each assembled valve for signs of severe fouling, corrosion attack, broken components, etc. Valves that appear to be in good condition should then be "leak" tested with Varsol (head-of-liquid test). All valves passing visual inspection and "leak" tests should be reinstalled for additional service.

Procedure CI-8

On Clark HS and HT horizontal compressors, remove crosshead slippers and connecting rod bearings for possible regrooving.

Procedure CI-9

Inspect force-feed lubricator. See that all pumps operate properly. Replace defective pumps. Adjust rate.

Procedure CI-10

Regrout compressor (not to be done at scheduled intervals but only as necessary).

Procedure CI-11

Clean intercooler.

Procedure CI-12

Check setting of unloaders and reset as necessary.

Procedure CI-13

Check cylinder alignment. This includes inspection of supports and adjust-ments necessary to minimize vertical and lateral movement of cylinders.

Compression Equipment Safe Operating Procedures

Compression equipment is one of the most important elements of an operating plant. A malfunction may damage the equipment in a relatively short time. The information that follows is intended as a guide to safe practices in preparing operating instructions for compressors. (NOTE: The more detailed procedures included for the screw compressors indicate the type of information needed for any compressor system.)

General

The following general procedures should be used in connection with the operation of all compressors:

1. *Housekeeping.* Keep the area around compressors neat and clean. Clean up oil and water leaks promptly and eliminate the cause of the leakage. Keep stairways and walkways at compressors free from tools, oil containers, rags, and other obstructions.
2. *Instrumentation and Protective Devices.* Periodically check calibration and setting of instrumentation and protective devices, such as thermo-meters, pressure gauges, alarms, trips, relief valves, and standby oil pumps. Incorrect pressure or temperature readings can lead to serious troubles. Serious failures can occur if a protective device does not operate.
3. *Purging.* Compressors handling toxic or flammable gases should be isolated from the process system by blinds or double block valves and bleeds and then completely purged with inert gas before being opened for maintenance. Before returning these compressors to service, all air should be purged from the machine and adjacent piping with inert gas. Per-manently connected purge piping should be avoided unless blinded when not in use.

4. *Liquid Slugs.* Inlet and interstage piping, including knockout drums, pulsation bottles, and intercoolers, should be checked for liquid before starting a compressor. Traps and drainers should be maintained in proper working condition. A slug of liquid between the piston and cylinder head would cause the same destruction that a piece of iron or steel would.

5. *Observation.* Operators must learn to detect changes in sound, appearance, and vibration of compressors, as well as changes in operating temperatures, pressures, capacity, etc., and to relate these to performance. Changes in any of these characteristics are often the first indication of trouble. Early detection of abnormal conditions followed by the proper corrective action can prevent many serious accidents.

Reciprocating Compressors

Each reciprocating compressor consists of one or more cylinders. Operating instructions should indicate allowable compression ratios across each compressor stage. If this ratio exceeds the amount designed in by the manufacturer, an excessive rod loading occurs, and rod breakage can result. Also, the maximum pressure and discharge temperature for each cylinder should be defined. For example, the compression ratio for air compressors is usually limited to four in order to avoid temperatures greater than 350°F. (A nonlubricated compressor is preferred for air service. An alternate is to use Celulube.) A higher temperature may result in an explosion in a lubricated system. The points in regard to compressor operation that require particular attention are the following:

1. *Compressor Valves.* The suction and discharge valves in a compressor cylinder actually are check valves. On many compressors the suction and discharge valves are interchangeable and in some cases reversible. Thus mistakes can be made if proper precautions are not taken. If a suction valve is installed in a discharge port, the compressed gas cannot get out of the cylinder through the valve and dangerously high pressures may build up, particularly in high-pressure cylinders with only one discharge valve at each end. In fact, the cylinder may fail, the piston may break, or the drive system may be damaged. It is preferable that the compressor be designed so that the valves cannot be improperly installed. However, where this has not been done, each valve installation should be checked in place before the valve cover is put on. This can be done by pushing the valve strip, channel, plate, or whatever the mechanism may be with a pencil. If the valve is installed as a suction valve, the pencil will depress the plate, channel, or strip and this can be readily detected. If the valve is installed as a discharge

valve, the plate or other mechanism will be firm and no spring action will be felt. For identity purposes it is desirable to have the compressor cylinder painted to indicate the location of suction and discharge valves.

Valve breakage is the most frequent cause of compressor shutdowns. Operators, in making their rounds of the compressors, should check all of the valves and know the signs that indicate a bad valve. These are:

The valve is hotter than usual.

The cylinder discharge temperature is higher than normal.

The cylinder capacity is lower than normal.

The cylinder discharge pressure is lower than normal (if this pressure is not independently controlled).

The cylinder suction pressure is higher than normal (if this pressure is not independently controlled).

The interstage pressures are abnormal on multistage compressors.

A loose valve usually makes a clattering noise. If this condition is detected, the machine should be brought down as soon as possible to prevent possible breaking up of the valve such that it will drop into the cylinder.

2. *Start-up and Loading.* Compressors that have been shut down for extended periods or for maintenance should be barred over at least one revolution before starting to insure that there is no mechanical interference within the compressor. Also, before starting, the piping system should be checked for proper setting of valves and controls. Generally, reciprocating compressors should be started with no load, i.e., with the by-pass or vent valves open to avoid overloading the driver. To avoid excess pressures and overheating, make sure that the discharge valve is at least slightly open before the by-pass valve is completely closed.

Where multiple cylinders are used with separate vents, or by-passes are used in series, it is important that they be brought up to pressure in sequence starting with the low-pressure stage.

3. *Vibration.* Since reciprocating compressors have pistons and crossheads moving back and forth, some degree of vibration always exists. This vibration, even though small, can cause small auxiliary lines to fail. Small lines should be anchored and should be reinforced with welded gusset plates and connections to the equipment. All piping should be inspected periodically for cracks, particularly at flanges. The use of threaded connections on process piping between the compressor and the first valve should be avoided.

4. *Tail Rods.* High-pressure cylinders may be equipped with tail rods on the piston to balance the piston rod loading. There have been a number of

serious failures and tragic accidents where the tail rod has broken near the piston and the free end was "blown" out of the cylinder like a projectile. This allows the high-pressure gas to escape freely. To minimize the possibilities of such accidents, tail rods should be inspected by magnaflux and reflectoscope before installation and at periodic intervals thereafter. Tail rods should be rifle drilled to permit better inspection. Since tail rods often are surface hardened, care should be exercised during maintenance to avoid surface cracking. This, in fact, applies to all surface-hardened compressor rods. Also, tail rods should be enclosed in a steel housing or "catcher" of sufficient strength to stop the tail rod if it should fail and also to contain the high-pressure gas. The housing must be vented sufficiently at the outboard end to avoid compression of gas in the catcher during normal operation.

5. *Packing Leakage.*   Piston rods of reciprocating compressors are equipped with special packing rings to minimize gas leakage out of the cylinder. Generally, these packing rings require constant lubrication and, for high-pressure services, special cooling facilities. Since some leakage will always exist, a vent line should be provided to carry the leakage to a safe location. If the packing is allowed to deteriorate badly, the vent system may not carry away all of the leakage and it will escape around the compressor rod. The packing should be checked periodically for leakage, lubrication, and high temperature.

6. *Gaskets.*   Reciprocating compressor cylinders use either gaskets or metal-to-metal joints between the various pieces. Most of these joints are subjected to pressure variations during each stroke of the piston while the machine is operating. Therefore, care should be exercised when reassembling these components to insure that the proper gasket is used, that the surfaces are in good condition, and that the bolts or studs are tightened properly, particularly on high-pressure cylinders. Generally, it is poor practice to tighten a joint when the compressor is operating.

7. *Discharge Temperature.*   Each compressor cylinder should be equipped with a temperature indicator at the inlet and outlet, and certain critical services should have a high-temperature alarm at the outlet of the cylinder. These temperatures should be noted and recorded on a scheduled basis. Changes in temperature should be investigated. An increase in discharge temperature can indicate a valve failure, increased internal leakage past the piston rings, an increased compression ratio, a change in gas composition, or a loss of cooling water. The cause of high discharge temperature should be corrected promptly to avoid serious damage. High discharge temperatures, over 350°F, can be dangerous in lubricated air cylinders since a fire, explosion, or detonation may result.

8. *Lubrication.*   Compressor cylinders and packing require constant lubri-

cation to avoid rapid wear and heating unless they are of a special design. The lubricant is usually furnished by a forced-feed lubricator unit mounted on the compressor. These lubricator units should be checked frequently for proper operation, feed rate, and oil level in the reservoir. Excessive oil feed rates should be avoided. On high-pressure service, extreme care should be taken to obtain a tight seal where oil lines pass through water jackets to assure that gas from a cylinder does not leak along an oil line into the cooling water system.

9. *Isolating and Purging for Maintenance.* To perform maintenance work safely on a compressor while the rest of the plant continues to operate, it is necessary to isolate the compressor from the rest of the system. This can be done by using single block valves and blinds or by using double block valves with a bleed valve in between. (To avoid the necessity of blinding, the cost of double block valves can usually be justified.) The use of a single block valve without a blind is not recommended because of the strong possibility that some leakage will occur through the valve.

Where the compressor suction and discharge lines are equipped with double block valves with bleed in between, the following procedure for isolating the compressor from the rest of the plant and preparing it for maintenance should be used. First, the compressor should be shut down. Then the double block valves should be closed and the bleeder in between should be opened and checked to make sure that it is not plugged. This can be readily determined by observing whether or not a small amount of gas flows out when it is opened. If the bleeder system is plugged, it should be unplugged using a special packing gland and a threaded rod tail. The compressor and the piping inside the double blocks should be depressured to the vent system. Following depressuring, the compressor cylinder to be worked on should be purged with an inert gas before it is opened. A suitable gas tester should be used to determine that the purging is complete (in the case of synthesis gas the gas tester should be calibrated for hydrogen and should also have flame arrestors designed for hydrogen). The compressor cylinder can then be opened for maintenance. It is also important to have some sort of hold card or lockout system on the compressor driver to prevent anyone from starting the compressor before the maintenance work is complete.

If single block valves are employed, the following procedure should be used. First, the block valves should be closed tightly. The compressor and piping inside the block valves should be depressured to the vent system. Then, at each block valve, a blind should be installed in the valve flange nearest to the compressor. The compressor cylinder to be worked on should then be purged using an inert gas before it is opened. From this point on, the procedure should be the same as with the system having double block valves with bleed in between.

When the maintenance work is completed, the air must be purged from the compressor cylinder and piping before the machine is returned to service. The cylinder is pressured and depressured with inert gas until the released gas contains less than 1% oxygen.

Gas Engines

Gas engines are often used as compressor drivers. The main safety precautions in their use are as follows:

1. *Starting Air System.* Numerous explosions have occurred in starting air lines to gas engines. Investigations have shown that faulty or leaking air check valves at the power cylinder probably caused most of these accidents. Therefore, these valves should be checked regularly and maintained in good condition. Also, excessive quantities of lube oil should be kept out of the starting air lines. The starting air header on an engine should be vented during normal operation.

2. *Shutdown Devices.* Gas engines preferably should be equipped with shutdown devices which shut off the fuel as well as de-energize the ignition system. If engines are not so equipped, the fuel and ignition should be shut off manually immediately after the engine stops. If the fuel is not shut off, unburned fuel will collect in the engine and in the exhaust system. If this happens and the system is not purged before restarting the engine, the fuel can explode in the exhaust system. It is good practice to purge the exhaust system with air by cranking the engine with the fuel and ignition shut off, regardless of the shutdown arrangement.

3. *Crankcase Explosions.* Explosions in crankcases of gas engines can be serious, and operators should be familiar with available preventive measures. These explosions result from ignition of a combustible mixture of oil or gas and air in the crankcase during operation. Since it is difficult, if not impossible, to eliminate all sources of ignition, crankcases are ventilated, or purged with an inert gas, and/or equipped with crankcase relief devices. Regardless of the precautions taken, the crankcase oil should be checked regularly for dilution, and excessive power piston blow-by should be eliminated. The crankcase should not be opened while the engine is hot.

Screw Compressors

This is a typical detailed operating guide. Other machines need similar data.

General

Functioning of the screw compressor is essentially the same as that of a reciprocating compressor. That is to say, the screw-type functions against any pressure prescribed by the network pressure or by a closed shut-off device, delivering a volume that is practically constant.

The following examples are given to show the operator the results of not operating the unit correctly and the lack of proper safety devices.

1. *Shut off valves in suction and discharge lines should be opened completely.* If the operator forgets to fully open the suction valve, it will result in less than normal initial suction pressure. This results in an abnormal heating of the rotors. Owing to the increased heat expansion, the rotors may make metal-to-metal contact which may result in a ruined compressor. A clogged strainer or filter will have the same effect.

    If the operator forgets to open the valve in the discharge line, the pressure upstream is greatly increased. In this case the increased pressure ratio may cause thrust bearing failure and overheating, leading to rotor seizure.

2. *The main shut off valve in the cooling water must be opened.* In a two-stage unit a failure of cooling water supply to the intercooler will cause increased heat expansion and consequently a damaging contact between the rotors of the first and second stages.

3. *Suction drum and suction and interstage lines must be drained before start-up.* A collection of liquid here will cause a hard-water hammer to wreck the compressor.

4. *Oil level of the oil console must be checked.* A low oil level may cause the oil pump to suck air. This turns oil into a foam with the pressure rising and falling. The bearings get insufficient lubrication, which can cause damage to the bearings.

5. *Water valves for case cooling are to be opened.* If the water cooling of the case fails to function for any length of time, the water in the case evaporates, thus causing an expansion of the case walls. This may cause the rotors to touch the casing at the gussets. When the cooling water is dirty, some deposits may be found at the underside opposite the water inlet. In most cases such deposits may be found by merely touching the casing during operating and finding a hot spot in the case wall. Deposits must be removed. The outlet temperature of the case cooling water is to be adjusted to about 104/122°F.

All of the above-mentioned possibilities of causing damage are minimized by automatic safety devices. Thus special attention should be given to the upkeep and maintenance of these devices.

Cold Start-Up (after long downtime)

1. Commission utilities.
   a. Air—instruments; blow down each filter regulator.
   b. Steam—main oil pump driver supply and exhaust, reservoir heating, steam tracing.
   c. Cooling water—coolers.
   d. Electrical—motor drives for pumps, conditioner, and/or centrifuge, electrical instruments, alarm power, immersion heaters for reservoir.
2. Fill oil reservoir and check quality of contents.
   a. Drain water, if any, from reservoir.
   b. Note action of reservoir.
   c. Establish nitrogen flow to oil reservoir.
   d. Heat oil to minimum operating temperature.
3. Charge accumulators to 35 lb/in.^{2}.
4. Check piping lineup in oil circuitry.
   a. Close high point vents and low point drains.
   b. Open block valves around control valves.
   c. Close bypass valves around control valves.
   d. Exercise air-operated control valves by manual loading.
   e. Establish oil loop circulation with primary oil pump.
      (1) Vent vapor from high points in piping, cooler, and filter.
      (2) Fill, vent, and switch filter/cooler path.
      (3) Purge air from pressure switch leads where possible.
      (4) Check that loop pressures are as expected.
         (a) Pump discharge pressure versus setting for auxiliary pump autostart.
         (b) Filter pump should be in clean condition.
         (c) Header pressure versus low oil pressure trip setting.
         (d) Accumulator level, where visible, corresponds to design value.
   f. Establish flow through the auxiliary lube oil pump.
      (1) Check that system pressure control handles flow from both pumps.
      (2) Shut down turbine pump.
      (3) Check header pressures—should be the same as for main pump.

   g. Check autostart capability of auxiliaries.
     (1) Establish flow with main pump.
     (2) Vent pressure at main pump discharge.
     (3) Assure that switchovers are accomplished without unit trip.
     (4) Note that "standby pump running" alarm is lit.
     (5) Shut down auxiliary pump.
   h. Miscellaneous protective instrument checks.
     (1) Trip on unit low lube oil pressure.
     (2) Check proper setting of bearing alarms.
   i. Check system for leaks.
     (1) Filters, flanges, valve stems, seals.
     (2) Check for casing oil accumulation.
   j. Test switchover valving for filter/cooling paths—should not actuate unit trip or start auxiliary pump.
   k. Commission seal gas flows and check for correct values.
   l. Commission oil purifier equipment.
   m. After main oil pump has run for a while:
     (1) Check for flow and temperature at all oil-using points.
     (2) Adjust cooler oil outlet temperature design value.
   n. With a strap wrench, turn gear input shaft clockwise when facing gear about two turns to make sure compressor rotors aren't stuck. (Motor should be locked out.)
   o. Press panel alarm to check alarm light and check for proper panel indication.
   p. Check for proper position of bypass unloading valve (should be open).
   q. Line up compressor circuit in accordance with process operating instructions.
   r. "Bump" motor to observe if motor rotation is correct.
   s. Start compressor.
     (1) Note time for unloading valve to close.
     (2) Listen for unusual noises and quickly check suction, interstage, and discharge pressures and temperatures.
   t. Start flushing oil pump and check counter to determine whether that pump is producing a white-oil flow.
   u. Observe machine closely for about 30 min. for any unusual symptoms and correct or shut down if correction isn't feasible.
   v. Mark time and date of start-up on bearing temperature monitor chart and take set of reading of process temperatures and pressures for log.

Restart (machine has been shut down for less than 12 h with the lube oil and oil purifier in operation and no major work has been performed on the machine)

1. Drain suction drum and suction and interstage piping.
2. Check pressures and temperatures of lube oil system and operate auxiliary pump by bleeding pressure switch. Auxiliary pump should start with no alarm generation.
3. Perform items n through v of Cold Start-Up list except item q may be omitted.

Shutdown Procedure

1. If possible, check one of the trip circuits on shutdown. For example, check the low-oil pressure shutdown by bleeding pressure off the pressure switch.
2. When motor is tripped, observe to determine if unloading valve opens before motor stops.
3. Also, check compressor stop to make sure back-rotation doesn't occur. Back-rotation indicates a stuck check valve in the discharge line.
4. Shut down white-oil injection pump.
5. If the compressor is to be down for a short time, leave the oil and seal system in commission.
   a. Before shutting off the oil system:
      Assure that cases are sufficiently cool.
      Check proper function of auxiliary pump.
      Check that reservoir level is safe to accept rundown without overflow.
      Shut down main lube oil pump by bleeding.
      Block supply and exhaust steam.
   b. Isolation:
      Block all utilities except air, and then only if the air system is being decommissioned.
      Keep condensate out of the reservoir by maintaining oil-purifier operation and N_2 bleed into reservoir.

Operations Monitoring via Routine Checks

1. Instrument checks twice each shift.
   a. Lube oil:
      Header pressure and temperature.
      Filter pump.
      Cooler pump.
      Reservoir level.

        Watch pump running.

        Oil purifier operating.

        User oil drain temperature and visual flow verification.

    b.   Compressor suction, interstage and discharge temperatures and pressures.

    c.   Bearing temperatures.

2. Periodic operating checks.

    a.   Drain reservoir of any water accumulation weekly.

    b.   Sample oil condition monthly.

    c.   Where possible, check trip function of switches monthly (requires test connection to bridge actual trip action).

Maintenance

General

Appropriate maintenance of all components will prevent damage and prolong the working life of any screw compressor installation.

In particular, all safety devices should be tested for proper functioning. The lube oil should be checked regularly for proper lubricity.

The type of gas, the quality of cooling water, and the surrounding atmosphere are important factors for determining the frequency of maintenance work. The maintenance schedule should be timed accordingly.

Inspection periods are also recommended in addition to the regular maintenance schedule. Under ideal operating conditions, a normal inspection should be performed every 8,000 h of operation and a major inspection made every 16,000 h of operation.

Major inspection includes the disassembling of the compressors. In general, it is recommended that this work be done with the assistance of the vendor's service personnel.

Spare parts used during inspection and maintenance should be reordered immediately.

Regular Maintenance Routine

Regular maintenance routine should start with observation of the machine. All deviations from normal, pressure and temperature in particular, should always be regarded as alarm signals.

The following data should be observed and recorded regularly:

1. Inlet pressure (each stage)
2. Inlet temperature (each stage)
3. Discharge pressure (each stage)
4. Discharge temperature (each stage)
5. Cooling water temperatures
6. Oil temperature
7. Oil pressure

Additional points of inspection to be checked and maintained regularly are:

1. All safety devices and instrumentation for proper setting and operation.
2. Unloading devised for proper functioning and leaks.
3. Check valves in air or gas piping for free movement and proper seating. A leaking discharge valve could run the compressor in reverse rotation at shutdown. This could lead to severe damage or destruction of rotors and bearings.
4. Safety valves for leaks and proper pressure settings. Leaking safety valves on multiple stage compressors can influence temperatures and rates of compression.
5. Intake filter for cleanliness and pressure drop.
6. Oil filter.
7. Oil level and proper lubricity of oil. Insure oil specifications are maintained.
8. Oil reservoir. Residue accumulation should be removed occasionally. If babbitt is found, all bearings must be removed and checked.
9. Oil supply lines, cleanliness of oil orifices, and tightness of all pipe connections.
10. Oil cooler. Maintain an oil temperature out of cooler between 100 and 110°F. Do not operate at an oil temperature below 90°F or above 130°F.
11. Bearing housings and cooling jackets for undue hot spots.
12. Intercoolers and aftercoolers. Check efficiency of coolers by noting inlet and outlet temperatures and by taking pressure drop across cooler. Check efficiency and proper function of moisture separators and drains.
13. Suction and interstage piping is to be checked for corrosion and cleaned as required. Rust in this piping will flake off and be carried with the air or gas into the compressor. This rust is hard, and a grinding process will take place in the sealing faces of the rotors. Serious damage can be done if the pieces are thicker than the clearance between the rotors.

14. Undue noise, check gears.
15. Vibration, check coupling alignment.

Normal Inspection (~8,000 h)

1. Check all items under regular maintenance routine.
2. Remove discharge and suction piping and silencers as required to inspect rotors through suction and discharge ports. Remove covers on suction and discharge ends of the compressor and the covers on the cooling jacket.
3. Clean cooling water jacket.
4. Make a visual inspection of the rotors through the discharge port. If necessary, flush with cleaning solvent, draining through the bottom port.
5. Check profile clearance on rotors and clearance between rotors and case bore.
6. Push rotors toward the suction side and check axial gap between rotor face and case on the discharge side.
7. Push rotors toward discharge side and check clearance.
8. Check oil return housings for babbitt. If babbitt is found, all bearings must be removed and checked.
9. Check radial bearing clearance. This is accomplished by lifting both ends of the rotor vertically as far as they will go and measuring the distance lifted with a dial indicator. The permissible clearances are as follows:

Shaft diameter (in.)	1.575	1.165	2.755	3.150	3.940	4.920	6.300	7.480
Maximum clearance (in.)	0.004	0.005	0.007	0.0075	0.008	0.009	0.0095	0.0095

10. Make visual inspection of timing gear teeth, and check tightness of cap screws and slotted nuts.
11. Remove gear box cover, and check teeth for proper wear and backlash between bull gear and pinions. The following are maximum backlash figures, based on bearing diameters. For shaft diameters up to 2.500 in. the maximum backlash can be $2\frac{1}{2}\%$ of the bearing diameter; for shafts over 2.500 in. in diameter the figure is 2%.
12. Check oil lines and clean orifices.
13. Check torsion bars for alignment, straightness, and tightness of all bolts and nuts.
14. Check main and auxiliary oil pump, including pump drives.
15. Check low-speed coupling and alignment of driver and gear.
16. Replace all covers.

Major Inspection ($\sim$ 24,000 h)

After $\sim$ 2 to 3 yr of operation, it is recommended that a complete inspection of all compressor parts be made.

New seals and bearings should be procured prior to disassembly.

Materials of Construction

JOSEPH A. CAMERON
FRANK M. DANOWSKI, Jr.
MARILYN E. WEIGHTMAN

Introduction

Selection of materials to use in rotating machinery must begin with careful
study and appraisal of expected operating conditions. In the discussion which
follows, attention is focused primarily on centrifugal compressors, but similar
consideration is required for other types of turbomachinery, such as axial
compressors and steam and gas turbines.

In addition to the expected conditions of operation, it is necessary to take
variations into account, both temporary and of long duration. Further,
machinery purchased for intermittent operation requires careful study because
it usually experiences the most severe kind of service.

Environmental factors must also be considered:

1. The operating stresses of rotating parts, determined by the design, size, and
   speed of rotation, must be known to select materials having sufficient
   strength.

2. The operating stresses of the casing and other stationary machine elements must also be known. For these parts the stress is determined primarily by design, size, and internal pressure.
3. The gas to be handled by the compressor must be evaluated for aggressiveness with respect to the materials of construction. If high corrosion rates will occur with ordinary materials, alternate materials must be selected, even though substantial additional cost may be incurred.
4. The expected temperature of operation must be determined. Temperature is a factor in all of the above: it affects the rate at which corrosion reactions proceed, affects the strength of materials, and influences susceptibility to brittle failure.

Properties Which Require Consideration

Virtually all of the properties of materials must be considered in evaluating materials for use in centrifugal compressors. Not all properties are important in every application, but various combinations must be evaluated in each instance, depending on service requirements. These properties include:

1. Tensile properties at room temperature and at actual operating temperature
2. Modulus of elasticity
3. Coefficient of thermal expansion
4. Susceptibility to brittle failure (toughness)
5. Damping
6. Fatigue strength
7. Thermal conductivity
8. Specific heat
9. Hardenability
10. Weldability
11. Corrosion resistance
12. Erosion resistance
13. System compatibility
14. Polymeric degradation

Omitted from the above list are long-time, high-temperature properties such as creep or stress rupture. Operation of centrifugal compressors at temperatures high enough to require consideration of these properties is very infrequent; such properties are much more commonly encountered in turbines.

Aspects to Be Considered

It would be impractical in one discussion to attempt to discuss all of the considerations involved in selecting materials for centrifugal compressors. A few of the most important are:

1. Impeller fabrication
2. Impeller materials
3. Coatings
4. Elastomers
5. Stress corrosion cracking
6. Sulfide stress cracking
7. Hydrogen embrittlement
8. Brittle failure
9. Dimensional stability

Impeller Fabrication

Types of Impellers

The heart of a centrifugal compressor is its rotating element; therefore, a great deal of attention is given to the fabrication of impellers. Fully closed impellers with solid back wall and cover, as shown in Fig. 1, are the most common type in multistage compressors. They present more difficult fabricating problems than the open or semiclosed type of impeller, more commonly used in single-stage machines (Fig. 2). For this reason, attention is focused primarily on fully closed impellers.

In most cases, centrifugal compressor impellers are fabricated from materials that are considered difficult to weld, but satisfactory techniques have been developed. Thousands of impellers have been fabricated successfully; in fact, the history of successful impeller welding dates back to the late 1930s.

Through the years there have been changes and improvements in materials and processes. Probably the most significant advance was the advent of low hydrogen materials and processes. Prior to this development, one of the most difficult problems had been that of delayed cracking. Impellers that appeared sound immediately after welding would occasionally crack seriously after standing at room temperature for periods ranging from a few hours to a few days. Also, these earlier impeller forgings were usually made from acid open hearth steel with relatively high phosphorus and sulfur contents which

FIG. 1. Typical fabricated closed impellers.

FIG. 2. Typical fabricated open impeller.

increased the cracking probability and lowered the weldability. In recent years this practice has been changed. Lower impurity, basic electric furnace steels are now used, and weldability has been improved [1].

Methods of Fabrication

Several methods of fabrications are shown in Fig. 3. Number 1 illustrates the most common method of fabrication: blades are fillet welded to the hub and cover. Welding may be done by manual arc (SMAW), gas shielded metallic arc (GMAW), or submerged arc welding (SAW) procedures. These welds may be readily inspected by magnetic particle methods on magnetic materials, and liquid penetrant methods on nonmagnetic materials.

This method of fabrication is the least costly welding procedure, and requires the least capital equipment cost. Its chief disadvantages are poor accessibility when the blade height (distance between hub and cover) is low or the backward lean angle of the blades is high, and intrusion of the fillet welds on the aerodynamic passage.

Number 2 is similar to Number 1 except that the welds are full penetration. This is a more costly welding procedure which seems to offer few, if any, advantages. The same welding procedures may be used as in Number 1.

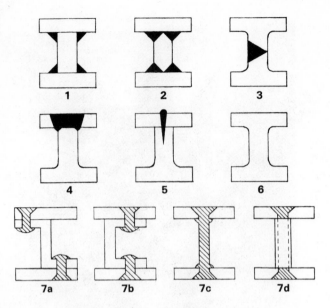

FIG. 3. Methods of impeller construction.

Theoretically, it would be possible to reduce the fillet size, thereby increasing the area of the aerodynamic passage, but in practice this reduction is small.

Number 3 has a partial blade machined integral with the hub and cover and a butt weld down the middle. This procedure has been used successfully, but is more difficult than it appears, especially with blades having high backward lean angles. There is also some difficulty in achieving a satisfactory weld contour around the leading edge of the blades.

Number 4, slot welding, is used primarily as an alternate to fillet welding in applications where the blade height is too small for accessibility, or the backward lean angle is too high to permit conventional fillet welding from the periphery of the impeller. This weld may be deposited by gas tungsten arc welding (GTAW), GMAW, SMAW, SAW, or a combination of these. As compared with fillet welding, there is a considerable increase in the amount of weld metal that must be deposited, and therefore an increase in welding time. Compensating for this to some extent is the lesser intrusion of the weld into the aerodynamic channel.

The blades may be attached to hub or cover by fillet welding, or they may be cast or milled integrally. This subassembly can then be welded to the slotted component. The slots may be placed on either the hub or cover.

Number 5 shows one possible configuration for electron beam welding a hub or cover to a blade. The chief problem, in this case, is the possible extension of the interface between the blade and hub or cover from which a crack can propagate into the weld on cooling, as shown in Fig. 4. Other joint configurations are possible which may eliminate fabrication and inspection difficulties. However, the possible advantages of this technique seem to be negated by high capital equipment costs, low equipment utilization, and the ability to fabricate a comparable product using more conventional techniques.

Number 6 illustrates a construction without welding where the impeller is either made from a one-piece casting or machined from a solid forging, possibly by electrodischarge machining. In one-piece castings the tooling cost is very high because of the varying blade heights and impeller cover eye openings. Electrodischarge machining is a relatively slow process that produces what is known as a "recast layer" on the surface of the part, Fig. 5. Studies have shown that this recast layer can reduce the fatigue strength of the part by as much as 60% unless it is removed by careful grinding [2].

Number 7 shows various configurations of riveted blades which have been used. Riveted parts are less strong than welded ones, and there is a potential source of fatigue cracking due to stress concentrations around the rivet holes. Numbers 7a and 7b have rivet heads which protrude into the gas passage, detracting from its aerodynamic cleanliness. If there is any dirt in the gas, it can collect and cause balance or corrosion problems.

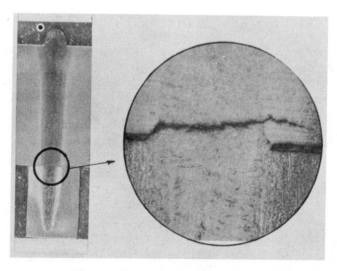

FIG. 4.   Electron beam weld showing interface cracking (5× and 50×).

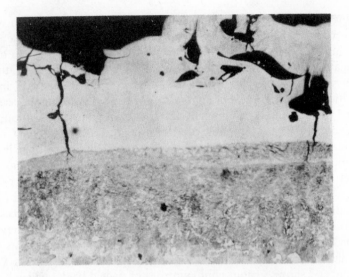

FIG. 5.   Recast layer from electrodischarge machining (250×).

Impeller Materials

Usual Materials

Impeller disk and cover forgings are usually made from an alloy steel such as AISI 4140 or 4340. The final selection of the particular alloy grade for each application depends on both the strength level desired and the size of the impeller. AISI 4140 is satisfactory for most applications; AISI 4340 is used for larger impellers or higher strengths because of its greater hardenability and resistance to tempering. Its greater resistance to tempering makes it possible to achieve higher yield strength while maintaining a relatively high tempering temperature. The usual minimum permissible tempering temperature is 1100°F, which makes possible a good balance between strength, toughness, and low internal stress. Chemical analyses and tensile properties of these and other impeller materials are listed in Tables 1 and 2.

Forgings of AISI 4140 and 4340 are heat treated to the desired strength level, Table 2. After machining and welding, the impellers are heat treated to temper the heat affected zone in the base metal adjacent to the welds. The base metal hardness varies with the yield strength required, but is frequently about Rockwell C-26. In the heat affected zone immediately after welding, the

TABLE 1 Chemical Analyses of Impeller Materials

Iron-base alloys	Carbon	Manganese	Silicon	Chromium	Molybdenum	Nickel	Copper	Columbium
AISI 4140	0.38–0.43	0.75–1.00	0.20–0.35	0.80–1.10	0.15–0.25	—	—	—
AISI 4340	0.38–0.43	0.60–0.80	0.20–0.35	0.70–0.90	0.20–0.30	1.65–2.00	—	—
AISI 410	0.15 max	1.00 max	1.00 max	11.50–13.50	0.50 max	0.50 max	—	—
AISI 304	0.08 max	2.00 max	1.00 max	18.00–20.00	—	8.00–12.00	—	—
AISI 304L	0.03 max	2.00 max	1.00 max	18.00–20.00	—	8.00–12.00	—	—
AISI 316	0.08 max	2.00 max	1.00 max	16.00–18.00	2.00–3.00	10.00–14.00	—	—
AISI 316L	0.03 max	2.00 max	1.00 max	16.00–18.00	2.00–3.00	10.00–14.00	—	—
Armco 17-4PH	0.07 max	1.00 max	1.00 max	15.50–17.50	—	3.00–5.00	3.00–5.00	0.15–0.45

Nickel-base alloys	Carbon	Manganese	Chromium	Molybdenum	Nickel	Copper	Columbium and Tantalum	Aluminum	Titanium	Iron
Monel K-500	0.25 max	1.50 max	—	—	63.00–70.00	Balance	—	2.00–4.00	0.25–1.00	2.00 max
Inconel 625	0.10 max	0.50 max	20.00–23.00	8.00–10.00	Balance	—	3.15–4.15	0.40 max	0.40 max	5.00 max

Aluminum-base alloys	Copper	Silicon	Iron	Manganese	Magnesium	Zinc	Chromium	Titanium
2025	3.90–5.00	0.50–1.20	1.00 max	0.40–1.20	0.05 max	0.25 max	0.10 max	0.15 max
C355	1.00–1.50	4.50–5.50	0.20 max	0.10 max	0.40–0.60	0.10 max	—	0.20 max

TABLE 2 Mechanical Properties of Impeller Materials

Material	Ultimate Tensile Strength (lb/in.²), min.	0.2% Yield Strength (lb/in.²)	% Elongation, min.	% Reduction in Area, min.	BHN
Iron-base alloys					
AISI 4140	100,000	80,000–90,000	16	50	212–235
AISI 4140	115,000	95,000–120,000	16	50	269–321
AISI 4340	120,000	105,000–120,000	17	43	248–302
AISI 4340	125,000	115,000–135,000	16	50	269–321
AISI 410	100,000	80,000–90,000	14	40	212–235
AISI 304	75,000	30,000 min	40	50	205 max
AISI 304L	75,000	30,000 min	40	50	187 max
AISI 316	75,000	30,000 min	40	50	217 max
AISI 316L	65,000	25,000 min	40	50	207 max
Armco 17-4PH	130,000	105,000 min	16	50	277–341
Nickel-base alloys					
Monel K-500	130,000	90,000 min	20	—	250 min
Inconel 625	120,000	60,000 min	30	—	—
Aluminum-base alloys					
2025T6	52,000	32,000 min	10	—	100 min
C355T62	44,000	33,000 min	2.5	—	—

hardness can be as high as Rockwell C-45 or 50. After tempering at 1100°F, this value falls to Rockwell C-28 to 30.

The use of more complex postweld heat treatments will be discussed later, but when a simple tempering treatment is employed, the temperature cannot be higher than 1100°F because this is the minimum tempering temperature for impeller hub and cover forgings. A postweld heat treatment at a temperature above the minimum tempering temperature for the forgings could result in a loss of strength in the hub or cover base material.

Special Materials

Where more corrosion resistance is needed than is available from ordinary alloy steels, such as in some chemical and petrochemical process gas compressors containing moisture and corrosive gases, AISI Type 410 stainless steel, which contains about 12% chromium (Table 1), is usually the first step. Weldability of Type 410 is comparable to that of the alloy steels.

A further improvement in corrosion resistance for more aggressive environments can be obtained through the use of one of the precipitation hardening stainless steels, such as Armco 17-4PH or 15-5PH. These grades, while appreciably more expensive, have the advantages of further improved corrosion resistance and much improved weldability as compared with Type 410 stainless steel. The precipitation hardening grades also have higher strength; for this reason they are sometimes used where the corrosion resistance of Type 410 would be adequate.

Austenitic stainless steels, such as Types 304, 304L, 316, and 316L (Table 2), are occasionally used in low tip speed impellers because their corrosion resistance is attractive in certain environments. However, their yield strength is only about 30,000 lb/in.^{2}, imposing a severe limit on tip speed and precluding their use for most applications. Types 304L and 316L have been used, or proposed for use, in single-stage compressors with 100% hydrogen sulfide. This gas is reportedly dry during normal operation, but may be wet during process upsets as well as during start-up and shutdown.

A further deterrent to the use of austenitic stainless steel impellers is that their coefficient of thermal expansion is about 50% higher than that of alloy steel shafts. The most common rotor construction consists of impellers shrunk on carbon or alloy steel shafts. If the temperature excursion during operation is sufficient, the differential growth of the shaft and impeller may cause the impeller to come loose on the shaft. A heavy shrink fit cannot be used to offset this problem because of the low yield strength of the material.

The temperature excursions in single-stage compressors are usually such that the differential coefficient of expansion can be handled, and these materials have given considerably improved resistance to such gases as sulfur dioxide.

Another material which has been used for centrifugal compressor impellers is Monel K500. It has been used primarily for exposure to halogen gases in the moist condition, and the record is generally good, although corrosion rates are sometimes high enough to require occasional replacement. It has also been used because of its resistance to sparking in oxygen compressors.

In the early 1950s there were several stress corrosion failures in hydrogen reformer compressors, but these were attributed to improper heat treatment resulting in a high level of internal stress. Since that time, many impellers have been fabricated using improved heat treating practices with no further difficulties. With Monel K500, as with the austenitic stainless steels, allowance must be made for its high coefficient of thermal expansion. However, its yield

FIG. 6.   Titanium compressor components for wet chlorine service.

strength is considerably higher than that of the austenitic stainless steels, making it possible to use heavier shrink fits.

Compression of halogen gases, especially chlorine, presents an interesting problem. If the chlorine is truly dry and the temperature is kept below about 275°F, standard materials may be used satisfactorily [3]. In a few instances, compressors for wet chlorine service have been built completely from commercially pure titanium. Their cast impellers were of the open-inducer type. The casing volute casting had a cleaned weight of 1800 lb, Fig. 6. All of the castings were vacuum cast in graphite molds.

While titanium works exceptionally well with wet chlorine, it cannot be used with dry chlorine because, in the event of a rub which would generate an elevated temperature, titanium is pyrophoric. This characteristic can be suppressed by adding moisture to the chlorine; the generally accepted minimum moisture content is 0.015%.

Inconel 625 has also been used as the impeller material in a compressor handling a gas high in chlorine. While its yield strength is not as high as that of the alloy steels, it is well above that of the standard austenitic steel compositions.

Aluminum alloy compressor impellers have been used in great numbers, but in a limited range of applications. Some higher strength aluminum alloys are becoming available, and the applications may be expected to increase. Limitations on their use have been (1) decrease in strength above approximately 300°F, and (2) difficulties in fabricating closed impellers from the high strength compositions.

Integral cast closed and semiclosed aluminum alloy impellers have been used extensively in diesel turbochargers and refrigeration compressors, and in some centrifugal compressors. Aluminum alloy impellers are generally not used for oxygen service because, in the event of a rub between the rotating and stationary parts causing the aluminum to reach a high temperature, there is a possibility of a Thermit reaction with oxides on the surface of the iron or steel parts.

Both aluminum and titanium are sometimes selected for centrifugal compressor impellers because of their low density, even in environments where corrosion will not occur with standard low alloy steels. This lower density causes a shift in the lateral critical speeds of a rotor which may be advantageous.

Protective Coatings

Stationary Components

The practicality of coating stationary components depends upon the accessibility of the surface for cleaning and coating. Single-stage compressor components, such as inlet ducting and back plates, are usually of simple geometry, permitting easy abrasive blasting and subsequent coating, Fig. 7. Inlets to multistage compressors and volutes of single-stage compressors are markedly more complex. Most casings, such as single-stage compressor volutes, Fig. 8, are marginal candidates for coating. Guide vanes in larger machines usually have good accessibility, but smaller machines may have extremely limited accessibility. Efficient functional design of interstage diaphragms (which redirect the gas flow to the next stage of compression, Fig. 9) results in gas passages that are very difficult to coat.

FIG. 7.   Single-stage compressor inlet duct suitable for coating.

Rotating Components

Rotating members such as impellers may require coatings to provide an antistick or "lubricated" surface or, in sensitive processes, to reduce the area exposed to the fluid medium. In one case, when a chemical process was reported to be sensitive to iron, all internal surfaces (including the impeller) of three single-stage compressors were coated with 0.001 to 0.002 in. of electroless nickel. Service to date has been satisfactory.

The most common reason for coating impellers is corrosion resistance. Steel compressor impellers have been coated to prevent erosion and corrosion, but only infrequently. Limited use has been made of organic coatings such as phenolic and epoxy resins, or their modifications. Chlorotrifluoroethylene and polyvinylidene fluoride resins, Kel-F (3M Co.) and Kynar (Pennwalt Corp.) have also had successful applications.

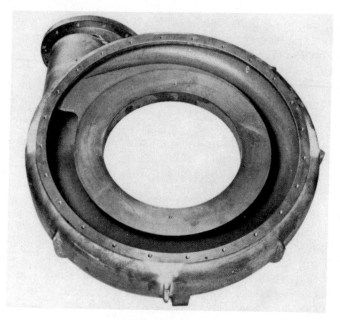

FIG. 8. Single-stage compressor volute whose internal surfaces cannot be cleaned or coated.

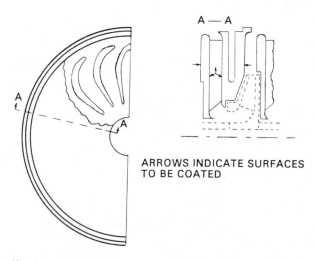

ARROWS INDICATE SURFACES
TO BE COATED

FIG. 9. Centrifugal compressor diaphragm schematic indicating functional surfaces requiring protective coatings.

To justify coating an impeller, the expense of both the coating procedure and the attendant subsequent modifications of manufacturing procedures must be considerably less than that of a material compatible with the environment.

With coated impellers, attention must be paid to:

1. Premachining to accommodate coating thickness where close dimensional tolerances are required.
2. A careful precoating surface inspection when a casting is involved.
3. Balancing both prior to and after coating.
4. Surface preparation prior to coating.
5. Protection of coating during transfer and assembly to avoid damage.

Consideration of the above items should provide a successful service life of from 1 to 2 years. At a scheduled overhaul, a prudent service procedure would be to install a spare coated rotor. (It is highly recommended that spares be available for coated impellers.) The original may then be recoated, rebalanced, etc. and be available for reinstallation at the next overhaul.

Check balancing of the impeller poses two problems. First, it entails handling under circumstances in which individual padded box protection can no longer prevent coating damage. Second, when it is not possible to grind away metal to achieve the required balance, it is necessary to spot remove the coating. The total surface area of coating removal must be kept to a minimum. Although subsequently returned to the best possible condition—a procedure that varies with each of the coating materials—the recoated area may be slightly less resistant to corrosive media than the original.

The condition of the cast or fabricated surfaces can jeopardize or enhance the use of a coating. Removal of blind hole porosity from all surfaces to be coated is mandatory, because the coating cycle will trap air in these holes if they are allowed to remain. Bubbling of the coating will occur during the heat curing cycle, interrupting continuity of the protective film.

Standard construction fabricated impellers (Fig. 3, Number 1) have a similar problem. Since the welds are not full penetration, their periphery must be completely seal welded prior to coating to assure coating continuity. This configuration can be satisfactorily coated, provided that sharp edges are blended and all other welds are continuous.

The required abrasive blasted surface preparation, such as SSPC (Steel Structures Painting Council) [4], Surface Preparation 10, or Surface Preparation 5, cannot be performed on small tip opening impellers because they are not large enough to accommodate cleaning and subsequent spraying of the coating.

Similarly, impeller configurations which prevent continuous welds are not good candidates, since coatings will not bridge gaps resulting from intermittent welds. Also, coatings will not bridge the sleeve to lashing wire gap in open impellers of Fig. 2 type construction. In these cases, corrosion-resistant materials should be selected.

Aluminum alloy impellers are almost universally anodized. They are sometimes conventionally anodized, but are more commonly hard anodized to obtain a thicker and harder coating that provides protection from both corrosion and erosion. In environments where particulate build-up may be a problem, the somewhat porous hard coat surface may be subsequently filled with TFE (tetrafluoroethylene) polymer.

External Surfaces

Special coating of external compressor surfaces does not involve an accessibility problem. However, unique surface preparation considerations do exist. For example, casings which ultimately house rotating elements cannot be abrasive blasted after rotor installation due to the possibility of damage to bearing and seal components. Thus casings requiring primers such as the inorganic zinc silicates must be coated before assembling the rotor in the casing.

The exterior of the casing is abrasive blasted to SSPC SP-10 (Near White). Within a 4-h maximum elapsed time period, it is then coated with a material such as self-curing inorganic zinc silicate having a nominal thickness of 0.003 in. With the cured primer intact, the casing then proceeds through the normal manufacturing channels.

Coatings requiring abrasive blast preparation cannot be applied to parts made of thin sheet metal. Since the cleaning procedure will distort metal less than $\frac{1}{8}$ in. thick, standard coupling housings, oil filter jackets, and equipment lagging cannot be adequately prepared. Either the metal thickness must be increased or the material selection changed to a corrosion-resistant grade.

In many cases, equipment jacketing constructed from 3XX series stainless sheet can cost less than increased thickness carbon steel plus the surface preparation and subsequent coating application.

Elastomeric Seal Selection

The life expectancy of nonmetallic materials should preferably equal that of metallic components; industry has aimed for a useful life of 3 years for nonmetallics such as elastomers. Accordingly, these components should be located in easily accessible positions to facilitate replacement during periodic inspections and overhauls.

In most compressor applications, critical elastomeric seals are in the form of "O" rings. (An "O" ring is a continuous gasket having an inside diameter and a width of cross section with tolerances established for both dimensions, Fig. 10 [5, 6].) They are basically static, accommodate slight lateral movement, and have either axial or radial squeeze.

Choice of an "O" ring material is clear-cut when the machine handles a single fluid. For example, a dry nitrogen gas compressor operating at 450°F discharge would have silicone "O" rings for static seals. However, simple applications such as this seldom occur, and most elastomeric seal choices are compromises. The decision involves weighing fluid media against service temperature and pressure requirements. Once the controlling criterion is established, trade-offs on minor criteria are made. Finally, the best commercially available material is selected for the specific service. Several examples of controlling criteria are:

1. Fluoroelastomers are not compatible with ammonia, ketones, esters, ethers, or amines.
2. Nitriles are not compatible with aromatic hydrocarbons.
3. Natural rubber and EPT (ethylene propylene terpolymer) are seriously degraded by conventional petroleum products.
4. Elastomeric seals must be individually selected for many synthetic lubricating and hydraulic fluids.
5. Nitrile is only usable to 212°F maximum.
6. Most fluorocarbon compositions are limited to 400°F maximum.
7. Standard silicone elastomers are limited to 450°F maximum.
8. Published data on temperature maxima are usually based on relatively short test exposures. Considering the long exposure time in service, it is often necessary to reduce published values by as much as 100°F.
9. "O" rings should not be used in high-speed rotary seals. Frictional heat will cause a tightening of the elastomer, which in turn generates more frictional heat. This becomes a self-feeding cycle that will destroy the seal in a short time (Gow-Joule effect).

Prior service experience in the same or a related environment is another important consideration in the selection of an "O" ring material.

Temperature limitations of common elastomeric materials (Table 3) allow matching elastomers to the operating conditions of the equipment.

Chemical resistance compatibility presents a somewhat more complex problem. Pure gases are the only straightforward applications. Mixtures may contain corrosive constituents, such as aromatic, aliphatic, or halogenated hydrocarbons. The elastomers in gas seals are exposed to the lubricant in

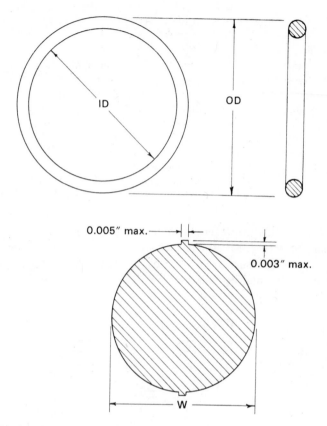

FIG. 10. Geometry noting conventional nomenclature and flash tolerances.

addition to the compressor fluid medium. A typical standard application chart is shown in Table 4. Publications distributed by various polymer manufacturers will usually contain references to services not listed in the table.

It is important to utilize standard ARP (Aircraft Recommended Practice) "O" ring sizes if at all possible because the cost of producing nonstandard size molds is prohibitive in most nonrepetitive applications. A 5% circumferential stretch is permitted in pressures less than 1500 lb/in.² gauge. Above this pressure, no stretch is permitted. Thus proper preplanning and design will result in a standard "O" ring size in most equipment. Suggested groove design data can be found in publications of various manufacturers.

Service pressures higher than 1500 lb/in.² gauge with standard squeeze dimensioning require back-up rings to prevent the extrusion shown in Fig. 11.

TABLE 3   Temperature Ranges of Elastomeric "O" Rings in Air

Temperature Range (°F)	Preferred Elastomeric Material
−150 to −40	Low-temperature silicone (special)
−80 to +450	Standard silicone
−40 to +175	Chloroprene
−40 to +212	Nitrile
−40 to +400	Fluoroelastomer
+68 to +550	Perfluoroelastomer
+450 to +500	High-temperature silicone (special)

Back-up rings of 90 durometer elastomeric material or fluorocarbon polymer are normally used (Fig. 12). Single back-up rings placed on the side of the "O" ring opposite the pressure are acceptable in many applications when lack of space prevents the dual installation.

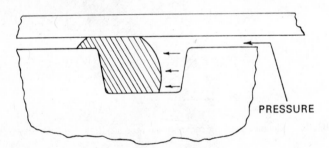

FIG. 11.   Schematic representing "O" ring extrusion under excessive pressure.

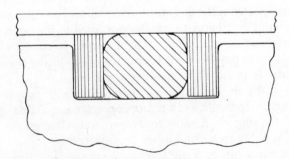

FIG. 12.   "O" ring back-up technique used to prevent extrusion.

TABLE 4 "O" Ring Application Chart

Media Groups	Temperature (°F), max	Elastomeric Material
Hydrogen and methane	212	Nitrile
	300	Fluoroelastomer
	400	Fluoroelastomer
Chlorine (dry), ethylene, and propylene	200	Fluoroelastomer
	300	Fluoroelastomer
	400	Fluoroelastomer
Air and nitrogen	212	Nitrile
	300	Fluoroelastomer
	400	Fluoroelastomer
	450	Silicone
Isobutane, butane, and propane	212	Nitrile
	400	Fluoroelastomer
Water, flue gas, ethylene, feedgas and recycle	212	Nitrile
	400	Fluoroelastomer
	450	Silicone
Refinery feedgas, benzene, toluene, xylene, coke oven gas, and many synthetic functional fluids	400	Fluoroelastomer
Petroleum base lubricants	250	Nitrile
	350	Fluoroelastomer
Wet refinery feedgas	400	Fluoroelastomer
	450	Silicone
Ammonia (gas)	175	Chloroprene
	212	Nitrile
	450	Silicone
Oxygen	200	Fluoroelastomer
	300	Fluoroelastomer
	400	Silicone

Two problem areas in the actual hardware assembly stage should be mentioned. The first is that slicing along a circumferential line sometimes occurs. This can most often be traced to improperly prepared "O" ring groove edges—they should be rounded to a minimum 0.005 in. radius. The second is that excess lubricant can contaminate the process gas. Lubrication of the elastomer facilitates installation and enhances service. A light film is sufficient.

The manufacturer's cure date of an elastomeric seal should not be more than 2 years old at the time of installation. The 5-year life expectancy of the seal thus includes the 2 years prior to its installation and 3 years of service, terminating at service overhaul. In modest applications at lower temperatures and in less aggressive environments, elastomers have experienced service lives much longer than 5 years.

Special Materials and Requirements

Although standard polymers have been discussed, mention should also be made of recent innovations in this field. For example, the low compression set fluorocarbons, available since 1970, have performed to expectations. A newer material is perfluoroelastomer, produced from fluoroelastomers and fluoropolymers. In many fluid media this material provides chemical and temperature compatibilities similar to TFE polymers while exhibiting characteristic elastomeric resilience. Applications in methyl ethyl ketone/toluene service at 350 and 400°F have been successful. Wet hydrogen sulfide service is suggested application. This blend represents an important extension in the elastomeric applications field.

Stress Corrosion Cracking

Stress corrosion cracking can be defined as the spontaneous failure of a metal resulting from the combined effects of stress and corrosion. This mechanism of failure is insidious in that there are virtually no tell-tale corrosion products or macroscopic reductions in the material thickness as fine intercrystalline or transcrystalline cracking develops.

If massive general corrosion is present, stress corrosion cracking does not usually occur. However, it may occur if sharp corrosion pits are present. If, in addition, cyclic stresses are present, the failure mode may be corrosion-induced fatigue. In this case, fatigue cracking starts at the sharp corrosion pits.

In order for cracking to occur, the following environmental requirements must be present:

1. An applied tensile stress
2. A corrosive medium
3. A concentration of oxygen

Cracking will only occur in a region of tensile stress. This stress can arise from several sources: it may be produced by operating conditions; residual stresses due to fabrication, machining, etc. may be of sufficient magnitude to cause failures; and, if residual stresses of low magnitude are combined with operating stresses, a level sufficient to cause cracking may be achieved.

The corrosive medium presents a complex problem. Logan [7] lists various alloys and media which have been reported to cause stress corrosion cracking. Alloys based on aluminum, magnesium, iron, copper and nickel are all susceptible given the proper combination of stress and corrosive medium. Stress corrosion has also been reported with mild steels in many different media. It is interesting to note that although nearly all of the engineering alloys are susceptible to stress corrosion cracking, pure metals do not fail by this mechanism.

It also appears that the oxygen concentration is important. For example, specimens of stainless steel exposed to a high concentration of chlorides coupled with a low concentration of oxygen did not fail. However, the specimens did fail when the oxygen concentration was increased (Fig. 13) [8].

The mitigation of stress corrosion seems, on the surface, to be fairly straightforward—eliminate tensile stresses and/or eliminate the corrosive environment. However, stresses due to operation cannot be eliminated. Residual stresses due to manufacturing may be minimized, but never eliminated and, while using materials in a fully annealed condition reduces stress corrosion, such use does not take advantage of alloys in their best condition of strength.

Some alleviation can be achieved by placing the material surfaces in compression by shot peening, rolling, etc. However, when pitting occurs due to corrosion, failure may be accelerated if the pits reach the region of high tensile stress below the peened surface layer.

Another method of combating stress corrosion cracking is to eliminate or change the environment. This may be done by changing the pH, eliminating the particular constituents causing the problem, or eliminating oxygen from the environment. However, doing any or a combination of these things is often not practical or feasible.

Although some work has been done on inhibitors and applied cathodic currents, there is usually only one practical solution to eliminate cracking: the material must be changed to one more resistant or immune to failure under a given set of service conditions. In some cases it is possible to use the same material if the strength level, heat treatment, or microstructure is changed.

The predominant problem to date in centrifugal compressors has been sulfide cracking rather than stress corrosion. The possible reasons for this are:

1. Steps taken to eliminate sulfide stress cracking may also eliminate stress corrosion cracking.

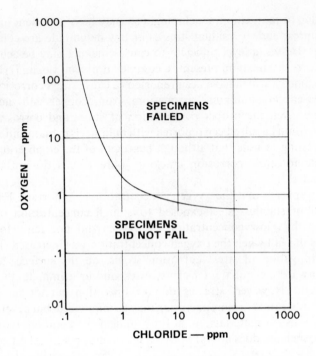

FIG. 13. Effect of dissolved oxygen and chlorides in high-purity water on susceptibility of stainless steels to stress corrosion cracking at 550°F [8].

2.   Normal gas compositions do not generally provide the necessary constituents.
3.   General corrosive attack may eliminate the formation of harmful pitting.

Consideration should be given to the possible build-up of contaminants, and consequent premature failure of compressor components, while the compressor is shut down.

Sulfide Stress Cracking

Background

In recent years, there has been a great deal of interest in the possibility of sulfide stress cracking in centrifugal compressor impellers. There is extensive literature on sulfide cracking in oil well casing materials [9] going back to about 1950. Incidents involving centrifugal compressors, however, have been rare, as reported by Kohut and McGuire [10] and Moller [11]. Nevertheless, the subject is one of serious concern because of the potential of a failure and subsequent loss of service of a vital link in the production chain of a chemical or petrochemical plant. Greer [12] recently presented an excellent paper describing the effects of more than a dozen variables on resistance to sulfide corrosion cracking.

Other investigations dealing specifically with sulfide cracking in compressor impeller materials have been reported by Scheminger, Ebert, and Aul [13] and by Keller and Cameron [14].

Essential Features

While it is impractical to review all of them in detail, the following conditions must be fulfilled for sulfide cracking to occur in susceptible materials:

1. Hydrogen sulfide must be present.
2. Water must be present in the liquid state.
3. The pH must be acid.
4. A tensile stress must be present.
5. Material must be in a susceptible metallurgical condition.

When all of the above conditions are present, sulfide cracking may occur with the passage of time.

Inhibition

It is frequently not possible to remove the hydrogen sulfide or moisture from the gas. Some interesting work has been reported on the prevention of sulfide cracking by the addition of inhibitors. As far as centrifugal compressors are concerned, the use of inhibitors does not seem to be widely practiced. The only variables that can be adjusted seem to be the condition of the material or the material itself.

Metallurgical Condition

Numerous investigations have shown that the optimum microstructure for resistance to sulfide cracking is tempered martensite resulting from heat

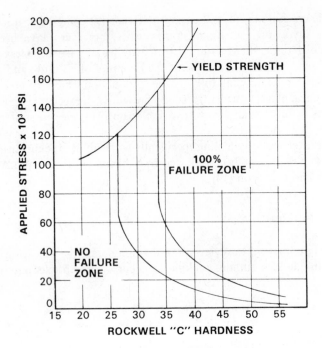

FIG. 14. Threshold stress for sulfide cracking as a function of hardness [15].

treatment by quenching and tempering. These studies have shown that alloy steels having a maximum yield strength of 90,000 lb/in.^{2} and a maximum hardness of Rockwell C-22 are not susceptible to sulfide cracking, even in the most aggressive environments. As the strength level increases above 100,000 to 110,000 lb/in.^{2}, the threshold stress required to produce sulfide cracking may actually decrease in very severe environments.

Warren and Beckman [15], for example, have reported that with a yield strength of 100,000 lb/in.^{2}, the threshold stress required to produce sulfide cracking approached the yield strength, while at a yield strength of 140,000 lb/in.^{2}, the threshold stress dropped below 30,000 lb/in.^{2}, as illustrated in Fig. 14. The test environment used by these authors involved a hydrogen sulfide water system at 104°F and 250 lb/in.^{2}.

Burns [16], using a different test environment, found the same inverse relationship of strength to resistance to sulfide cracking. His solution contained 5% sodium chloride and 0.5% acetic acid saturated with hydrogen sulfide. (This is the environment in a proposed standard method for evaluating resistance to sulfide cracking [17].) Burns tested several alloy steels, including AISI 4135 and

AISI 4135 modified with a molybdenum content of 0.82%. At Rockwell C-27 the two grades displayed comparable susceptibility to sulfide cracking. The work was reported to be continuing at lower hardnesses.

Grobner and associates [18] have reported improved resistance to sulfide cracking for AISI 4130 modified with 0.5 to 1.0% molybdenum, and AISI 4135 modified with 0.5 to 0.75% molybdenum and 0.035% columbium. The test solution was a 0.5% acetic acid solution saturated with hydrogen sulfide. Tests were conducted at room temperature. The improved performance was attributed to increased resistance to tempering, which required higher tempering temperatures to produce a given yield strength. Further, the increased hardenability assisted in achieving a microstructure of tempered martensite.

Effect of pH

Treseder and Swanson [20] have shown that the effect of pH on sulfide cracking is quite marked. For example, their experimental parameter (Sc) increased from 4 at pH 2 to nearly 14 at pH 5, as shown in Fig. 15. Sc is the stress value [(lb/in.^{2}) $\times$ 10^{-4}] corresponding to a 50% probability of failure.

The importance of pH was also shown in the recent work by Keller and Cameron [14]. Among the tests conducted in this work was a series on AISI 4140 which was quenched and tempered after welding to a base metal yield strength of 126,000 lb/in.^{2}. The test specimens were stressed to 80% of the yield strength and tests were conducted in triplicate. At pH 2.5, all three specimens failed, while at pH 4.2, none of the three failed. A similar series of tests was conducted on AISI 4140 which was quenched and tempered before welding to a base metal yield strength of 83,000 lb/in.^{2} with only a tempering treatment at 1100°F after welding. Again the test stress was 80% of the yield strength. At pH 2.5, three tests were conducted and all failed. At pH 4.2, there were eight failures in nine tests, but at pH 6.5 there were no failures in three tests. The test environment was room temperature water saturated with hydrogen sulfide, and the pH was controlled by addition of hydrochloric acid or ammonia water.

The acidity of the gas in a compressor is determined by condensing a sample of the gas and making a pH determination. When the concentration of hydrogen sulfide is sufficient to saturate the water, the pH value is about 4.3. If the hydrogen sulfide present is not sufficient to saturate the water, the pH will be higher. The pH may also be influenced by the presence of other constituents in the gas stream which are soluble in the water.

Hydrogen chloride, for example, is capable of depressing the pH to values well below 4.3 and accelerating attack. Other possible constituents could reduce the acidity and raise the pH, thus reducing the severity of attack. For these complex reasons it is difficult, if not impossible, to specify a threshold

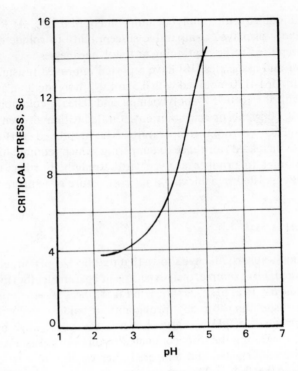

FIG. 15. Critical stress for sulfide cracking as a function of pH [20].

concentration of hydrogen sulfide in a gas below which sulfide corrosion cracking will not occur.

Effect of Temperature

A strong relationship between temperature and sulfide cracking has been reported by Hudgins [21] with time to failure increasing markedly from 75 to 150°F. Most centrifugal compressors with hydrogen sulfide in the gas operate at these or higher temperatures, particularly when the temperature rise through the machine is taken into account.

Service Experience

While the limits of 90,000 lb/in.^{2} yield strength and hardness of Rockwell C-22 were selected for very aggressive environments, experience suggests that the environment in centrifugal compressors is considerably less aggressive. There are many welded AISI 4140 and 4340 alloy steel impellers in service that were given a simple tempering treatment at 1100°F after fabrication and suffered no distress, although the environments are known to contain hydrogen sulfide.

Perhaps, in these cases, either the operating temperature of the compressor is above the dew point of the gas so that liquid moisture is not present, or the composition of the gas may be such that the pH is greater than seven.

Garwood [19] has cited other cases where satisfactory service has been obtained in exposure to environments containing hydrogen sulfide at hardness values of Rockwell C-28.

In many of these cases the hardness of the base metal is approximately Rockwell C-26, and the hardness of the heat affected zone adjacent to the welds is Rockwell C-28–30. In one case, as reported by Moller [11], in ethylene off-gas service there was cracking in the heat affected zone adjacent to some of the welds because of the omission of the 1100°F postweld heat treatment after repair welding. In this instance the hardness of the heat affected zone was Rockwell C-45–50. In the portions of the impellers which had not been repaired, and where the hardness of the heat affected zone was Rockwell C-30 or below, there was no cracking. It must be concluded that, in this instance, the severity of the environment was such that the common specification requirement of Rockwell C-22 maximum was not needed to prevent sulfide corrosion cracking.

Prevention

Compressor manufacturers are now equipped to perform a complete quench and temper heat treatment after fabrication by welding. When this is done, the heat affected zone disappears, as shown in Figs. 16 and 17, and the impeller base metal is of uniform hardness. With the use of such treatments, it appears safe to increase the maximum permissible yield strength up to at least 110,000 lb/in.^{2} [14].

In evaluating the maximum permissible yield strength, it is instructive to take into account the actual operating stress in centrifugal compressor impellers and its relationship to the yield strength. This stress increases with the square of the speed of rotation. According to American Petroleum Institute Standard 617, *Centrifugal Compressors for General Refinery Services*, it is required that impellers be overspeed tested at 115% of the maximum permissible continuous speed prior to assembly of a compressor rotor. The maximum continuous speed is, in turn, 105% of the design speed. Thus, at overspeed, the actual speed is 121% of design, and the stress is 145% of the stress at design speed. For these relationships, refer to Table 5.

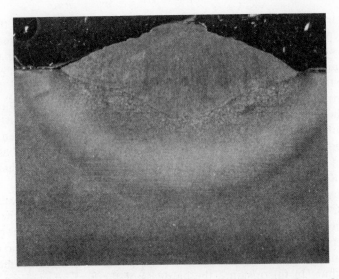

FIG. 16. AISI 4140 welded and tempered showing heat-affected zone (5×).

Assuming that the stress during overspeed testing approaches the minimum specified yield strength of the material, the stress at design speed would be only 69% of this minimum yield strength. At maximum continuous speed rating, the operating stress would be only slightly higher, still only 75% of the yield strength. These conservative stress values operate to reduce the risk of difficulty due to sulfide cracking.

The shafts of centrifugal compressors are made from heat treated alloy steels, usually the same grades, AISI 4140 and 4340, as in impellers. It is often required that their yield strength be over 90,000 lb/in.^{2} to accommodate the maximum applied stress under the coupling. At this location the shaft is not exposed to the process gas. The operating stress in the portion of the shaft inside the casing where it is exposed to the process gas is quite low—normally under 7000 lb/in.^{2}. For this reason, and taking into account data such as that previously cited from Warren and Beckman, the yield strength is usually not restricted.

Stationary parts, including the casing and diaphragms made from carbon steel and cast iron, meet the maximum strength and hardness requirements for immunity to sulfide cracking with margin to spare. Barrel-type compressors for very high pressure service are sometimes made from alloy steel forgings heat treated to yield strengths above 90,000 lb/in.^{2} and hardnesses over Rockwell C-

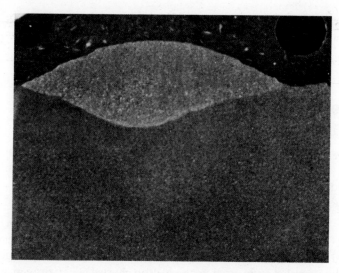

FIG. 17. AISI 4140 quenched and tempered after welding showing no heat-affected zone (5×).

TABLE 5 Impeller Stresses at Various Speeds of Rotation

Speed	Speed (%)	Stress (%)	Stress as % of Minimum Specified Yield Strength[a]
Design	100	100	69
Maximum continuous	105	110	75
Trip	115	130	90
Overspeed	121	145	100

[a]Percentages based on assumption that stress at overspeed testing approaches but does not equal the minimum specified yield strength. Actual conditions will reduce percentages by some factor.

22, but the operating environment is usually not one likely to give rise to sulfide stress cracking.

The National Association of Corrosion Engineers has recently (April 1978) published a new revision of Standard MR01-75, replacing an older document.

Material furnished to this standard may be expected to withstand exposure to very aggressive sulfide stress cracking environments. It is appropriate to repeat again that many actual service situations are of less than maximum aggressiveness, and some materials of higher strength are often acceptable in specific applications.

Hydrogen Environment Embrittlement

The effect of hydrogen in centrifugal compressors is another recent topic of interest. The hydrogen problem takes three different forms which have been described by Jewett et al. [22]:

1. *Hydrogen chemical reactions.* This has also been called hydrogen attack, and occurs at elevated temperatures—the minimum temperature is usually cited as about 425°F. Material selection for avoidance of hydrogen attack in steel is well documented by the Nelson charts [23]. Hydrogen attack has not been a serious problem in centrifugal compressors because high temperature and high hydrogen concentrations have not often occurred simultaneously. When the temperature has been high, the hydrogen content has been low and vice versa.
2. *Internal hydrogen embrittlement.* In this case delayed failure may occur, especially with high strength steels. Many examples are found during the manufacturing of such hardened and electroplated parts as springs, washers, and aircraft landing gear struts. Internal hydrogen was also the cause of flakes in forgings which were involved in the failure of several large turbine and generator rotor forgings some 20 years ago [24, 25]. The problem of flaking in large forgings has been eliminated with the use of vacuum degassed steel.
3. *Hydrogen environment embrittlement.* This is the aspect of the hydrogen problem which is of most significance in centrifugal compressors. It occurs while the metal is stressed in hydrogen. Material exposed to high pressure hydrogen, but subsequently stressed at room temperature, does not exhibit this embrittlement.

Features of Hydrogen Environment Embrittlement

An extensive literature has been developed on the last two forms of hydrogen embrittlement, including papers by Elsea and Fletcher [26], Walter and Chandler [27], Beck et al. [28], Cavett and Van Ness [29], and Steinman, Van Ness, and Ansell [30]. With all of the research, however, a 1969 NASA publication [31] stated that "There are considerable gaps in the data available; essentially, no thresholds have been established for specific influences such as pressure, temperature, and stress much less the influence of combined factors." To this list might be added effect of variations in gas composition.

There are several important distinctions between the last two types of hydrogen embrittlement. Internal hydrogen embrittlement does not show up immediately after charging the specimen, but will cause delayed failure if the specimen is held under a modest stress for a period of time. Hydrogen environment embrittlement, conversely, causes immediate embrittlement when the stress level exceeds the yield strength. Specimens usually fail during, or very shortly after loading, or not at all. In hydrogen environment embrittlement, failures start at the surface even in notched specimens, while in internal hydrogen embrittlement, cracks are initiated below the surface. Most of the time to failure in hydrogen environment embrittlement takes place in successive incubation periods. The incubation period before each step in crack formation is the time required for hydrogen to diffuse to the region of active crack progression.

Hydrogen environment embrittlement is sensitive to strain rate because it requires the movement of hydrogen in the presence of a stress gradient. The effect is at a maximum at low strain rate. Conventional impact tests do not indicate embrittlement because the strain rate is high and the time is too short to permit diffusion of hydrogen to the advancing crack tip. The degree of embrittlement is independent of holding time under pressure [27], even when the specimens are held under stress in high pressure (10,000 lb/in.^{2}) hydrogen.

Categories of Hydrogen Environment Embrittlement

Walter and Chandler [27] have separated metals into four categories according to the degree of embrittlement observed in high pressure hydrogen:

1. *Extreme embrittlement.* High strength steels and high strength nickel base alloys are in this category where embrittlement is characterized by a large decrease of notch strength and notched and unnotched ductility, and some decrease in unnotched strength in 10,000 lb/in.^{2} hydrogen. Metals in this category usually fail with one catastrophic crack which propagates into the specimen, leaving a thin shear lip around the periphery except at the site of crack initiation.

2. *Severe embrittlement.* The majority of metals tested were in this category

which includes most of the materials used in centrifugal compressors. Embrittlement is characterized by reduction of notch strength and notched and unnotched ductility, but little or no reduction of unnotched strength. Metals in this category usually fail with many surface cracks, some of which are quite deep.

3. *Slight embrittlement.* This category includes commercially pure titanium, copper, beryllium, and the austenitic stainless steels having an unstable austenite structure. Embrittlement is characterized by a small decrease of notch strength and notch ductility. Metals in this category exhibit numerous small, shallow, blunt cracks under low power magnification.

4. *Negligible embrittlement.* Materials in this category include aluminum alloys, stable austenitic steels, and copper. They fail in the same manner in hydrogen and in air with no reduction in strength or ductility due to hydrogen environment embrittlement. Surface cracking is not observed.

Effect of Variables

The effects of variables on hydrogen environment embrittlement have not been as fully defined as is desirable, presumably due to the extreme difficulty and cost of experimentation.

Hofmann and Rauls [33] have reported on the effect of temperature on normalized low carbon steel. In their work, embrittlement seemed to be most extreme near room temperature, decreasing as the temperature was raised or lowered. As shown in Table 6 [29], studies on AISI 4140 quenched and tempered to two different strength levels showed that the relative loss in strength due to hydrogen is approximately the same at 80 and 250°F. However, the high strength material (212,000 lb/in.^{2} yield strength and 228,000 lb/in.^{2} tensile strength) exhibited a greater reduction in its notched tensile strength than the low strength material (127,000 lb/in.^{2} yield strength and 135,000 lb/in.^{2} tensile strength).

Loginow and Phelps [32] have studied the effect of yield strength and hydrogen pressure. They determined values of critical stress intensity below which crack growth is arrested in hydrogen gas (K_H) and in air (K_{1X}), see Table 7. The 1-in. thick wedge opening loading specimens used in this work did not satisfy the constraint requirements necessary to obtain plane strain conditions for many of the steels. Thus K_H and K_{1X} were used rather than K_{1C}, which is the fracture toughness under conditions of plane strain. (For more detailed explanation of fracture toughness, see the section entitled "Low Temperature Operation—Brittle Failure.")

Elastic properties, including the yield strength of a material, are the same in hydrogen as in air. Some plastic deformation is required to initiate hydrogen environment embrittlement. For this reason it has been observed that varying the heat treatment of alloy steels produces different susceptibilities to embrittlement in a hydrogen atmosphere. Tempered martensitic structures, resulting from quenching and tempering, are less prone to embrittlement than structures produced by other heat treatments.

TABLE 6 Failure of Low- and High-Strength AISI 4140 in High-Pressure Hydrogen and Nitrogen

	Average Ultimate Strength (lb/in.^{2})[a]	% Reduction of Notched Tensile Strength Due to H_2
Low Strength, 4140		
80°F unnotched	135,000[b]	
80°F notched		
10,000 lb/in.^{2} N_2	241,000 (4)	
6,000 lb/in.^{2} H_2	207,000 (2)	14
10,000 lb/in.^{2} H_2	204,000 (3)	15
250°F Notched		
10,000 lb/in.^{2} N_2	221,000 (2)	
6,000 lb/in.^{2} H_2	207,000 (2)	6
10,000 lb/in.^{2} H_2	185,000 (1)	16
High Strength, 4140		
80°F unnotched	228,000 (2)	
80°F notched		
10,000 lb/in.^{2} N_2	362,000 (2)	
2,000 lb/in.^{2} H_2	135,000 (2)	63
6,000 lb/in.^{2} H_2	121,000 (4)	66
10,000 lb/in.^{2} H_2	89,000 (3)	75
250°F notched		
10,000 lb/in.^{2} N_2	274,000 (1)	
2,000 lb/in.^{2} H_2	92,000 (2)	63
6,000 lb/in.^{2} H_2	96,000 (4)	65
10,000 lb/in.^{2} H_2	82,000 (3)	70
350°F notched		
10,000 lb/in.^{2} H_2	103,000 (1)	

[a]Number of specimens shown in parentheses.
[b]Reported by the Bethlehem Steel Company.

The yield strength to tensile strength ratio is usually 0.85 or higher in quenched and tempered alloy steels. Other heat treatments, such as normalizing or normalizing and tempering, produce structures having lower strengths with a yield strength to tensile strength ratio as low as 0.5. Thus deformation begins at a lower fraction of the tensile strength in the lower strength materials with the result that a greater depression in tensile strength due to hydrogen environment embrittlement is observed.

The effect of purity of the hydrogen is most important. As little as 1% oxygen has been reported [33] to eliminate embrittlement in some alloy steels. Other impurities in the gas, which could combine with the metal to form a surface film presenting a barrier to the passage of hydrogen, would also be expected to be beneficial. Conversely, constituents in the gas stream which would be inert or which would attack the surface and prevent the formation of a barrier film would not have any beneficial effect, although possibly not a detrimental one. If oxide films are developed, but are subsequently ruptured by plastic deformation, they would be ineffective in retarding hydrogen environment embrittlement.

Some work has been done on protective coatings, but with limited success. The chief difficulties have been discontinuities or lack of adherence. Coatings are thin and could be damaged by a momentary internal rub between rotating and stationary compressor parts, or by a foreign material passing through the compressor in the gas stream, thus destroying the protective effect.

The pressure of the hydrogen gas is another important variable, at least as it affects the rapidity with which the gas can diffuse into an advancing crack. In higher pressures where the gas penetrates the crack more rapidly, conditions favoring the progress of hydrogen environment embrittlement at the crack tip are reached more rapidly. Consistent with this, it has been observed that fatigue strength is reduced in hydrogen environments, and that crack growth rates are increased over those observed in air.

Prevention

An important variable in determining the degree of hydrogen environment embrittlement is the yield strength of the material. In many cases, gas streams which are high in hydrogen also contain hydrogen sulfide and moisture. It has therefore become quite common to apply a maximum yield strength specification. Most commonly, the yield strength has been limited to 90,000 lb/in.^{2} maximum. Immunity to hydrogen environment embrittlement has been achieved simultaneously with immunity to sulfide stress cracking.

Most of the research on hydrogen environment embrittlement has been

TABLE 7 Values of K_H for Steels Exposed to Hydrogen at Various Pressures

Steel	Yield Strength $\times 10^3$ lb/in.^{2}	K_{1x} [$(10^3$ lb/in.$^2)\sqrt{\text{in.}}$]	K_H [$(10^3$ lb/in.$^2)\sqrt{\text{in.}}$][a]				
			3,000 lb/in.^{2}	6,000 lb/in.^{2}	9,000 lb/in.^{2}	10,000 lb/in.^{2}	14,000 lb/in.^{2}
Resistant steels							
Type 304	34	64				NCP-62	NCP-50
A516	42	76				NCP-75	NCP-81
A106	50	61					
HY-80	85	114				NCP-106	
Steels with moderate susceptibility							
A372 N and T	85	137	76	63	67	59	63
A372 Q and T—1100	87	128	64	50	64		40
4130 Q and T—1175	92	114	80	62	41	29	47
4145 Q and T—1100	97	139	66	61	50	55	28
A372 Q and T—900	101	110	—	—	—	50	—
4147 Q and T—1235	105	141	88	85	60	—	42
A517 Grade F	110	143	78	61	70	64	74
4147 Q and T—1200	113	144	112	37	41		27
Steels with appreciable susceptibility							
4147 Q and T—1170	126	146	35	27	22	—	21
HY-130	136	168	33	29	—	22	—
4145 Q and T—1050	153	104	20	17	—	—	—

[a]NCP = no crack propagation at indicated stress intensity levels.

conducted on materials having a yield strength of 180,000 lb/in.[2] or higher. When notched specimens have been tested, both strength and ductility have been substantially reduced. This embrittlement has been observed to decrease with decreasing strength, as reported by Cavett and Van Ness [29], by Steinman, Van Ness, and Ansell [30], and Deegan [34].

While the limit has not been precisely defined, it appears that yield strengths up to about 120,000 lb/in.[2] are acceptable in centrifugal compressors without undue risk of embrittlement. The limiting condition may be associated with the lesser fracture toughness of higher strength steels, and with the increased stress gradient at the base of a notch or crack in higher strength materials.

The relationship between the stress at operating speed and that during factory overspeed testing has already been reviewed. As in the case with sulfide cracking, this relationship provides a margin of safety in hydrogen environment embrittlement. During impeller overspeed testing, there may be some plastic deformation and some blunting of stress concentrations. Such testing is commonly carried out in a partial vacuum. Subsequently, operation at a lower stress in a more severe environment is less hazardous.

Low Temperature Operation—Brittle Failure

Background

When compressors are required to operate at subzero temperatures, consideration must be given to the problem of brittle failure. As temperatures fall, all materials become stronger and less ductile, and some materials become increasingly susceptible to brittle failures. Those materials which have this susceptibility include most of the materials commonly employed in compressor construction. In order for brittle failure to be initiated, there must be a stress exceeding the yield strength.

While the nominal stress is always well below the yield strength, it is not possible to design and manufacture any engineering structure without stress concentrations. It is from these sites that brittle failures propagate. Once initiated, brittle failures can propagate at very much lower stresses.

Such failures propagate with the speed of sound. This is the reason for reports of a loud noise accompanying brittle failures, such as those which occurred when welded ships broke in half during World War II. This had not been encountered previously in riveted ships because brittle failures usually did not propagate across riveted joints. Individual plates were replaced when they cracked. In the case of welded ships, the problem was not with the welding. However, a path for propagation across the joint existed, and catastrophic failures were encountered.

Criteria for Resistance to Brittle Fracture

There have been several changes in the criteria for susceptibility to brittle failure. In the work done immediately after World War II, the most commonly accepted criterion was the low temperature Charpy V notch impact test which used energy absorption to determine susceptibility. It had gradually come to be recognized that the previously used Charpy keyhole test, which also used energy absorption as the criterion, was not as definitive as desired. (See Fig. 18.) In the transition zone between ductile and brittle fracture, there is a spread in the Charpy keyhole values where the energy level obtained is unpredictable; the energy level obtained is more definitive with the Charpy V notch test.

Other tests have been used including the explosion bulge test, the drop weight test, and the drop weight tear test. These tests are more direct measurements of suceptibility to brittle failure. They suffer, however, from the disadvantages of requiring large size test specimens, expensive test setups, and test specimens not readily obtainable from some forms of material.

The ultimate criterion for resistance to brittle fracture seems to be fracture mechanics. A great deal of work has been done in this area starting in the 1950s, particularly with respect to ultrahigh strength materials and certain nonferrous alloys, such as aluminum and titanium. At its present state of development, fracture mechanics is not fully applicable to low alloy steels at the strength and temperature levels commonly encountered in centrifugal compressors.

The fracture mechanics approach takes into consideration that flaws exist in all materials and structures. The size of flaw that can be tolerated is a function of the material's fracture toughness and the applied stress level. A relatively large flaw may have little detrimental effect in a material with high fracture toughness, at ambient temperature, and with a low applied stress. However, a relatively small flaw, which may be difficult to detect by inspection techniques, may cause a catastrophic failure in a material with low fracture toughness, at low temperature, and with a high applied stress.

By proper analysis, using a fracture mechanics approach, a reasonable estimation may be made both of the possible consequences of an existing defect and of the maximum size of defect that can be tolerated for a given set of operating conditions.

The fracture toughness of a material is designated by K_{1C}, which is the critical stress intensity factor measured under plane strain conditions. Plane strain is a stress state that is characteristic of thick or bulky parts for which the stress adjacent to the flaw is triaxial tension. Materials with high fracture toughness have high values of K_{1C}.

For a given material, K_{1C} varies as a function of heat treatment, yield strength, and temperature. It decreases as temperature decreases and yield strength increases.

The basic expression in the application of fracture mechanics is

$$K_{1C} = \alpha\sigma\sqrt{\pi a}$$

where K_{1C} = critical stress intensity [(lb/in.^{2})$\sqrt{\text{in.}}$]
  α = a factor, usually between 1 and 2, which is a function of crack geometry
  σ = applied stress (lb/in.^{2})
  a = one-half the crack length (in.)

Knowing the values for K_{1C}, α, and σ would allow the calculation of the critical flaw size [35]. Values of K_{1C} and flaw geometry relationships are available.

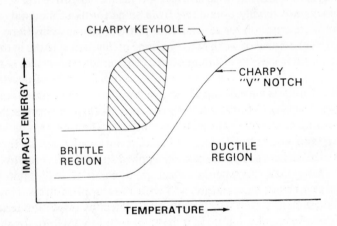

FIG. 18.   Charpy keyhole and V-notched impact energy as a function of temperature.

To make a realistic assessment of the risk of brittle failure, the following factors must be considered:

1. Temperature
2. Material thickness
3. Magnitude of stress and strain (both applied and residual)
4. Yield strength of the material
5. Size and acuity of the crack

6. Strain rate
7. Type of microstructure at the crack tip

Material thickness is important in that fracture toughness decreases as the thickness of a given material increases. This is partly due to microstructural variations, but mainly due to increasing effects of stress triaxiality. Above a certain thickness, the fracture toughness remains essentially constant. When this occurs, plane strain conditions exist and the fracture toughness value is a valid K_{1C} [36].

Fracture mechanics can also assist in cases where a structure is subjected to cyclic stressing. If a subcritical flaw is present, it may grow by fatigue to critical size. The relationship between the rate of crack growth (da/dN) and the change in the stress intensity factor (ΔK) is the most important contribution of fracture mechanics to fatigue design. The quantity da/dN is the growth rate of a flaw in microinches per cycle of stress reversal. When da/dN is plotted vs ΔK, the slope of the curve is very nearly the same for all steels. Depending on the strength level of the material and its toughness, the curve may be shifted to the left or right (Fig. 19) [37].

For a given combination of flaw geometry and component configuration, crack growth rate is a function of the instantaneous stress intensity at the crack tip. In turn, K is a function of the applied stress and crack length. Under cyclic loading, K increases because both the crack length and stress increase. As the number of cycles increases, K reaches a level characteristic of the material where unstable crack growth results in fracture.

This relationship will allow the calculation of either the initial allowable flaw size for a given stress and life or vice versa [38].

Materials and Fabricating Procedures

Casing materials for low temperature service generally require only modest strength, and there are several grades of both wrought and cast products.

For fabricated casing material, the following steels are available: fine grain carbon steels (ASTM A516 Grade 60) with satisfactory properties at temperatures to $-50°F$; $2\frac{1}{4}\%$ nickel steel (ASTM A203 Grade A) and carbon–manganese–silicon steel (ASTM A537 Class 1) to $-75°F$; and $3\frac{1}{2}\%$ nickel steel (ASTM A203 Grade E) to $-150°F$. The properties of these materials at low temperatures are sometimes enhanced by accelerated cooling after the austenitizing treatment. This is particularly true for heavier thicknesses.

In order to take advantage of these materials, it has been necessary to develop special welding procedures. No difficulty is encountered in meeting the

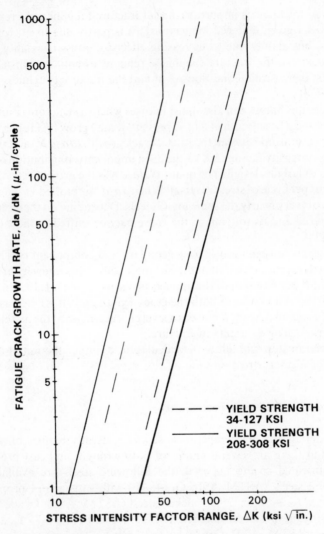

FIG. 19. Fatigue crack growth rate (da/dN) as a function of stress intensity factor range (ΔK) [37].

requirements at temperatures down to $-50°F$ using carbon steel filler metal for either manual metal arc or submerged arc welding. For temperatures below $-50°F$, the alloy steel filler metal and special fluxes needed for submerged arc welding are available. Representative data are shown in Tables 8 and 9 for tests on weld metal and heat affected zones, respectively.

Recently, it has been necessary to develop welding procedures suitable for fabrications at an operating temperature of $-160°F$. In this case the base metal used was a $3\frac{1}{2}\%$ nickel steel that was specifically qualified for $-160°F$ rather than the standard temperature of $-150°F$.

TABLE 8 Weld Metal Impact Strength[a]

Weld Metal	Method	Temperature (°F)	Impact Energy (ft · lb), Average	Lateral Expansion (mils), Average
E7018	Manual	-50	51	46
L61–860[b]	Sub arc	-50	23	21
E8018-C1	Manual	-75	42	34
L61-AXXX19S[b]	Sub arc	-75	53	46
Cryoweld 3[c]	Manual	-150	31	24
XW-19[d] and 709–5[e]	Sub arc	-150	18	11
6% nickel	Manual	-175	38	—
Cryoweld 3	Manual	-175	32	—

[a]In submerged arc welds, first grouping refers to wire designation, whereas second refers to flux.
[b]Lincoln Electric Co.
[c]Champion Commercial Industries
[d]Armco Steel Corp.
[e]Linde Division, Union Carbide Corp.

The manual metal arc welding posed no serious problems. Cryoweld 3 (Champion Commercial Industries) was the electrode chosen, since it had been used successfully to repair $4\frac{1}{2}\%$ nickel castings. This electrode has acceptable impact properties at $-175°F$. Some welding procedures that give $3\frac{1}{2}\%$ nickel weld deposits which are acceptable at $-150°F$ are marginal when the temperature is decreased to $-160°F$.

The use of special alloy wire and alloy containing fluxes makes it possible to qualify a submerged arc procedure for $-160°F$. The tests indicated that the weld metal could be used to approximately $-200°F$. However, this temperature

TABLE 9 Heat-Affected Zone Impact Strength

Base Metal	Method	Temperature (°F)	Heat-Affected Zone Impact Energy (ft · lb), Average	Heat-Affected Zone Lateral Expansion (mils), Average
A516 Grade 55	Manual	−50	119	79
A516 Grade 55	Sub arc	−50	49	45
A203 Grade A	Manual	−75	43	34
A203 Grade A	Sub arc	−75	53	44
A352 Grade LC3	Manual	−150	23	18
A203 Grade D	Sub arc	−160	33	24
$4\frac{1}{2}\%$ nickel casting	Manual	−175	21	

would require a change of base metal to either a 9% nickel or austenitic steel.

Repair welds in a $4\frac{1}{2}\%$ nickel material may be accomplished successfully by using a nominal 6% nickel steel manual metal arc electrode. This electrode is capable of developing sufficient impact strength at −175°F. Use of normal $3\frac{1}{2}\%$ nickel electrodes would be marginal at best, but Cryoweld 3 can be used successfully under most conditions. However, the use of 6% nickel electrodes allows a closer match to the base metal and represents a considerable reduction in cost as compared with the nickel-base electrodes.

Cast materials for casings are also available for service to −175°F using one of several nickel containing alloy steels with $2\frac{1}{4}\%$ nickel being satisfactory to −100°F, $3\frac{1}{2}\%$ nickel to −150°F, and $4\frac{1}{2}\%$ nickel to −175°F. Below −175°F, materials which have an austenitic structure must be used. Down to −320°F, conventional cast austenitic stainless steels, such as the cast equivalent to AISI Type 304, possess good fracture toughness. A modified 20% nickel cast iron has also been used for application at −260°F, and toughness values indicate that it would be acceptable to −320°F.

While conventional austenitic stainless steels such as Type 304 are more difficult to cast, and more expensive, they can be welded more readily. In addition, the properties of the base metal, heat-affected zone, and weld metal are all good at low temperatures. Modified 20% nickel cast iron has better foundry properties and requires fewer repair welds. There is more of a problem, however, with welding, especially at the heat-affected zone of the base metal. For this reason, welds in critical areas are not permitted, but noncritical locations are sometimes repaired.

In both materials, because of their austenitic structure, the coefficient of thermal expansion is about 50% greater than that of standard ferritic casings. This is not an insurmountable problem, but it must be taken into account, especially if the rotor is constructed of a material with a lower coefficient of expansion.

Rotors for operation at $-260°F$ have been fabricated from 9% nickel steel. This alloy has satisfactory resistance to brittle failure at temperatures at least as low as $-320°F$. Its coefficient of expansion is similar to that of the alloy steels, and thus much lower than that of austenitic casings. In the form of forgings and plates up to about 2 or 3 in. in thickness, the alloy has attractive properties, but it is not available in the heavier thicknesses required for casing flanges. Castings of 9% nickel steel are not used because, while they can be made, they do not have attractive properties.

Materials for Rotating Parts

Extensive work at the Naval Research Laboratory by Pellini and associates has shown that the energy absorption requirement for the prevention of brittle failure increases with the strength of the material. While a level of 12 or 15 ft · lb in the Charpy V notch test is sufficient for ordinary mild steel, the energy absorption requirement might be 30 ft · lb or more for quenched and tempered alloy steel [39].

Recognizing this problem, Gross and associates at Lehigh [40] and at U.S. Steel [41] proposed a criterion using a direct measure of notch ductility. They examined both the lateral contraction at the root of the notch and the lateral expansion on the compression side opposite the notch. They concluded that either would serve, but that lateral expansion was easier to measure than lateral contraction. For this reason, expansion was selected. Gross et al. proposed a criterion of 15 mils minimum lateral expansion for a wide range of strength levels. This has since been incorporated in the ASME Boiler and Pressure Vessel Code, Section 8, Division 1 for materials having a tensile strength greater than 95,000 lb/in.^{2}.

There is, however, still some difference of opinion on this subject, as indicated in a paper by Puzak and Lange [42] of the Naval Research Laboratory. This work demonstrates the superiority of other types of tests, especially the drop weight and drop weight tear tests. Unfortunately, these tests require test specimens which are difficult or impractical to obtain for some forms of material.

In the case of rotating elements, AISI 4330 or 4340 can be heat treated to give acceptable values using lateral expansion (per ASME Boiler and Pressure Vessel Code, Section 8, Division 1) for temperatures as low as $-150°F$ with a yield strength of 90,000 lb/in.^{2} or greater. At $-100°F$, the same impact test requirement can be met with a yield strength of about 105,000 lb/in.^{2}.

Some years ago, $3\frac{1}{2}$% nickel steels such as AISI 2340 were used for this application. More recently, use of AISI 4330, which is lower in nickel but higher in chromium and molybdenum, has become almost universal. The impact requirements can be met more readily with the 43XX series because of its increased hardenability and better response to heat treatment [43]. Below $-150°F$, other materials must be selected, such as the 5% nickel steel recently developed by Armco under the tradename Cryonic 5 or the 8 or 9% nickel steels.

As previously mentioned, 9% nickel steel has been used quite successfully at temperatures down to $-320°F$ in both rotating and stationary applications. For some applications, aluminum alloys have also been used successfully.

Stability

One of the most insidious problems that can occur in the manufacture and use of turbomachinery is dimensional instability. When instability is encountered during manufacture, it may be impossible to achieve required close tolerances on stationary parts and concentricity on shafts. Long-term instability which may occur in service can cause leakage at casing joints, or a rotor to go out of balance with resulting excessive vibration.

Casings

Instability leading to manufacturing difficulty can occur with centrifugal compressor casings, but in practice is not a major problem. Casings of either the cast or fabricated type are given a stress relief heat treatment during manufacture at about $1100°F$, a temperature which insures a low level of internal stress.

Generally, there is more difficulty with distortion during machining of castings than during machining of fabricated casings. The wall thickness of castings is tapered in order to achieve progressive solidification of the casting. As a result, the amount of metal removed in machining is not equally distributed. This tapered thickness is not encountered with fabricated casings. The amount of material removed in machining is more uniform and usually less than with castings. Further, castings may require substantial amounts of repair welding. Defects which are encountered during final machining, and which require welding, are particularly troublesome.

Long-term service instability of compressor casings is very seldom encountered. For this reason, literature in this area is nonexistent. This type of instability is more commonly encountered with turbines because of the higher operating temperature, greater temperature excursions in service, and higher thermal gradients. In addition, particularly with modern high-temperature, high-pressure turbines, the wall thickness of the casing is greater.

Several papers, including one by Reisinger and Scharp [44], have been published on the subject of steam turbine casing distortion and cracking due to thermal gradients. The use of slower heating and special design features in the casing avoids this problem.

Other internal stationary parts in centrifugal compressors, such as diaphragms, are most commonly made from cast iron. Occasionally, when higher strength is required, ductile iron or fabricated mild steel is used. In all cases the castings or fabrications are heat treated to produce a low level of internal stress, and difficulty with instability in service is virtually unknown.

Rotating Parts

Rotating parts for centrifugal compressors are, likewise, carefully treated in order to insure a low level of internal stress. Shafts made from quenched and tempered alloy steels, such as AISI 4140 or 4340, are tempered at a minimum temperature of 1100°F, and difficulty with achieving concentricity during manufacture is rarely encountered. As in the case of stationary parts, the problem of instability is a good deal more serious with turbines, and, again, published literature on this problem generally deals with them. For turbine rotors with integral disks, a vertical heat treatment is recommended. Further, such rotor forgings are tested for thermal stability in a special test apparatus. The rotor may be rotated at a speed of about 2 rev/min while being heated to a temperature of about 1000°F, with the eccentricity being measured by the use of dial indicators having long extension arms reaching into the test furnace.

Possibly the most common cause of true instability is deficiency in prior heat treatment. The result is that the material on one side of the shaft has a coefficient of thermal expansion different from that on the other side by about 1%. This is sufficient to cause a significant deflection when the shaft is heated to 800 to 1000°F. This and other possible causes of instability have been discussed at length in excellent papers by Barker and Jones [45] and Timo and Parent [46].

Summary

The considerations involved in selecting materials to use in centrifugal compressors are many and complex. Some of the most significant ones have been examined:

1. The physical and chemical properties which must be evaluated have been identified.
2. Impeller materials are usually alloy steels, such as AISI 4140 or 4340, with standard heat treatments. Special treatments are used for resistance to sulfide stress cracking. Other materials such as stainless steels, nickel alloys, and aluminum alloys are sometimes used for special applications.
3. The use of coatings for surface protection has been reviewed and applications for several coatings identified.
4. Information has been presented on the uses and limitations of elastomeric seal materials.
5. Sulfide stress cracking of compressor materials, especially impellers, has been a topic of much interest. With proper attention to materials and especially the condition of heat treatment, the risk of sulfide stress cracking is virtually eliminated.
6. In compressors handling gases consisting mainly of hydrogen there is a potential problem with hydrogen environment embrittlement. As in the case of sulfide stress cracking, solutions are available by proper selection of materials and treatment.
7. When compressors operate at low temperatures, there may be a risk of brittle failure due to lack of fracture toughness. Materials which will operate satisfactorily have been identified.

References

1. H. W. Mishler, R. E. Monroe, and P. J. Rieppel, "Studies of hot cracking in high strength weld metals," *Weld. Res. Suppl.*, 26, 1s–7s (January 1961)
2. J. Kahles, "Electrical Discharge Machining," in *Metals Handbook*, Vol. 3, 1967, pp. 227–233.
3. M. H. Brown, W. B. Delong, and J. R. Auld, "Corrosion by chlorine and by hydrogen chloride at high temperatures," *Ind. Eng. Chem.*, 39 (July 1947).
4. Steel Structures Painting Council, *Surface Preparation Specifications ANSI A159.1*, March 16, 1972.
5. Rubber Manufacturers Association Inc., *Rubber Products Handbook*, December 1970.
6. Rubber Manufacturers Association Inc., *Rubber Products Handbook*, "O" Ring Inspection Guide and Surface Imperfection Control, 1974.
7. H. L. Logan, *The Stress Corrosion of Metals*, Wiley, New York, 1966, pp. 5–7.
8. W. L. Williams and J. F. Eckel, *J. Am. Soc. Naval Eng.*, 68, 93 (1956).
9. L. W. Vollmer, "The behavior of steels in hydrogen sulfide environments," *Corrosion*, 14, 324t–328t (July 1958).

10. G. B. Kohut and W. J. McGuire, "Sulfide stress cracking causes failure of compressor components in refinery service," *Mater. Prot.*, 7, 17–22 (July 1968).
11. G. E. Moller, *Corrosion, Metallurgical, and Mechanical Experiences of Petroleum Refinery Compressors*, NACE Task Group T-8-1 Interim Report, 1968.
12. J. B. Greer, *Factors Affecting the Sulfide Stress Cracking Performance of High Strength Steel*, NACE Paper No. 55, 1973.
13. W. J. Scheminger, H. E. Ebert, and E. L. Aul, *Resistance of Some Standard Compressor Materials to Hydrogen Sulfide Stress Corrosion Cracking*, ASME Paper 71-Pet-25, 1971.
14. H. F. Keller and J. A. Cameron, *Laboratory Evaluation of Susceptibility to Sulfide Cracking*, Carrier Corporation, NACE Paper #99, 1974.
15. D. Warren and G. W. Beckman, "Sulfide corrosion cracking of high strength bolting material," *Corrosion*, 13, 631t–646t (October 1957).
16. D. S. Burns, *A Method for Evaluation of New Materials and Processes for Use in H_2S Service*, NACE Paper 75, 1975.
17. J. B. Greer, *Results of Interlaboratory Sulfide Stress Cracking Using the NACE T-17-9 Proposed Test Method*, NACE Paper 97, 1975.
18. P. J. Grobner, D. L. Sponseller, and W. W. Cias, "Development of higher strength H_2S resistant steels for oil field applications," *Mater. Perform.*, 14(6), 35–43 (1975).
19. G. L. Garwood, *Material Selection for Downhole and Surface Equipment for Sour Gas Condensate Wells*, NACE Paper 53, 1973.
20. R. S. Treseder and T. M. Swanson, "Factors in sulfide corrosion cracking of high strength steels," *Corrosion*, 24, 31–37 (1968).
21. C. M. Hudgins, *The Effect of Temperature on the Aqueous Sulfide Stress Cracking Behavior of an N80 Steel*, NACE Canadian Western Regional Conference, 1971.
22. R. P. Jewett, R. J. Walter, W. T. Chandler, and R. P. Fromberg, *Hydrogen Environment Embrittlement of Metals*, NASA CR2163, March 1973.
23. G. A. Nelson, *Action of Hydrogen on Steel at High Temperature and High Pressure*, Welding Research Council Bulletin 145, Section II, October 1969.
24. H. D. Emmert, *Investigation of Large Steam Turbine Spindle Failure*, ASME Paper 55A172, 1955.
25. G. Schabtach, A. W. Rankin, E. L. Fogleman, and D. H. Winne, *Report of the Investigation of Two Generator Rotor Fractures*, ASME Paper 55A208, 1955.
26. A. R. Elsea and E. E. Fletcher, *The Problem of Hydrogen in Steel*, DMIC Memorandum 180, October 1, 1963.
27. R. J. Walter and W. T. Chandler, *Effects of High Pressure Hydrogen on Metals at Ambient Temperature*, Final Report NASA Contract NAS19, 1969.
28. W. Beck, E. J. Jankowsky, and P. Fischer, *Hydrogen Stress Cracking of High Strength Steels*, Naval Air Development Center Report MA-7140.
29. R. H. Cavett and H. C. Van Ness, "Embrittlement of steel by high pressure hydrogen gas," *Weld. Res. Suppl.*, 28, 316s–319s (July 1963).
30. J. B. Steinman, H. C. Van Ness, and G. S. Ansell, "The effect of high pressure hydrogen upon the notch tensile strength and fracture mode of 4140 steel," *Weld. Res. Suppl.*, 30, 221s–224s (May 1965).

31. Anon, *Effects of Hydrogen on Metals*, NASA Tech Brief 69–10372, September 1969.
32. A. W. Loginow and E. H. Phelps, *Steels for Seamless Hydrogen Pressure Vessels*, ASME Paper 74-Pet-4; *Trans. ASME J. Eng. Ind.*, 97, 274–282 (February 1975).
33. W. Hofmann and W. Rauls, "Ductility of steel under the influence of high pressure hydrogen," *Weld. Res. Suppl.*, 30, 225s–230s (May 1965).
34. D. C. Deegan, *Exposure Tests in High Pressure Hydrogen Gas*, U.S. Steel Internal Report (September 19, 1967).
35. C. C. Osgood, *Mach. Des.*, 44, 58–64 (July 22, 1971).
36. K. G. Richards, *Brittle Fracture of Welded Structures*, The Welding Institute, Cambridge, England, 1971.
37. T. W. Crooker and E. A. Lange, *The Influence of Yield Strength and Fracture Toughness on Fatigue Design Procedures for Pressure Vessel Steels*, ASME 70-PVP-19, June 1970.
38. C. C. Osgood, *Mach. Des.*, 44, 88–94 (August 5, 1971).
39. P. P. Puzak and W. S. Pellini, "Evaluation of the significance of Charpy tests for quenched and tempered steels," *Weld. Res. Suppl.*, 21, 275s–290s (June 1956).
40. J. H. Gross and R. D. Stout, "Ductility and energy relations in Charpy tests of structural steels," *Weld. Res. Suppl.*, 22, 151s–159s (April 1957).
41. J. H. Gross, "The effect of strength and thickness on notch ductility," *Weld. Res. Suppl.*, 34, 441s–453s (October 1969).
42. P. P. Puzak and E. A. Lange, *Significance of Charpy V Test Parameters as Criteria for Quenched and Tempered Steels*, Naval Research Laboratory Report 7483, October 1972.
43. H. Schwartzbart and J. P. Sheehan, *Impact Properties of Quenched and Tempered Alloy Steels*, Armour Research Foundation of Illinois Institute of Technology—Project B0111, September 1955.
44. R. H. Reisinger and C. B. Scharp, *Thermal Stress Protection in Starting and Loading Boiler-Turbine-Generator Combinations*, ASME Paper 60PWR3, 1960.
45. A. Barker and F. W. Jones, *The Reversible Bending of Turbine Shafts with Temperature*, Institution of Mechanical Engineers (London), 1955.
46. D. P. Timo and D. F. Parent, *Thermal Distortion of Turbine Rotors*, ASME Paper 58A270, 1958.

Index

305

About the Authors

HEINZ P. BLOCH   is an Engineering Associate for Exxon Chemical Company, Baytown, Texas. He received his B.S. and M.S. degrees from the New Jersey Institute of Technology. Mr. Bloch is the author of over 25 publications covering a wide range of topics in mechanical engineering and holds four U.S. patents. He is a member of the American Society of Mechanical Engineers and American Society of Lubrication Engineers.

JOSEPH A. CAMERON   is Manager of Materials Engineering for the Elliott Company, Jeannette, Pennsylvania. He is the author of a number of papers on compressor and turbine materials. Mr. Cameron is a Fellow of the American Society for Metals and a member of the American Society of Mechanical Engineers, National Association of Corrosion Engineers, and American Foundrymen's Society.

FRANK M. DANOWSKI, JR.   is an engineer specializing in scanning electron microscopy and failure analysis for Elliott Company, Jeannette, Pennsylvania. He received his B.S. degree (1960) from Carnegie Institute of Technology. Mr. Danowski is a member of the American Society for Metals.

RALPH JAMES, JR.   was Chief Machinery Engineer for Exxon Chemical Company, Florham Park, New Jersey before his untimely death in 1979. Mr. James received the B.S. degree from the University of Texas and the M.S. degree from the University of Houston. He was an avid inventor, holding a number of patents for his work.

JUDSON S. SWEARINGEN is President of Rotoflow Corporation, Los Angeles. He received his M.S. (1930) and Ph.D. (1933) degrees from The University of Texas. Dr. Swearingen is the author of over 30 publications and holds more than 60 patents. He is a member of the American Chemical Society, American Society of Mechanical Engineers, National Academy of Engineering, American Institute of Chemical Engineers and corresponding (foreign) member of the National Academy of Engineering of Mexico.

MARILYN E. WEIGHTMAN is Senior Materials Engineer for Elliott Company, Jeannette, Pennsylvania. Ms. Weightman is a member of the American Chemical Society, American Society for Testing and Materials, and Society of Plastics Engineers.